The Role of Ethics and the Environment in Educational
Management in Higher Education A Case Study

Nirmal Chand

CONTENTS

Abbreviations
List of Tables
Acknowledgements

CHAPTER		Page No.
I	Introduction	1
II	Review of Literature	50
III	Higher Education in India	80
IV	Research Methodology	109
V	Analysis of Data and Presentation of Report	137
VI	Summary and Conclusions	167
	Bibliography	217
	Annexure	

Chapter – 1

Abstract

This chapter deals with the conceptual and theoretical aspects of the study covering education in general, with special reference to higher education in India, historical and philosophical aspects of ethics its dynamics and transformation to contemporary times and how it becomes vital in contributing to higher education and its management. In this process management and its methods were also reviewed very concisely. A brief review of education and its general principles and views of various authors are attempted. A history of philosophy and ethics up to contemporary time's vis-à-vis ethical schools and philosophers is presented very briefly.

INTRODUCTION

Andhra University, Visakhapatnam

"The only thing that interferes with my learning is my education."

_____ Albert Einstein

The need for education evolved and unfolded itself with the realisation of the fact that it is the key to the future and progress of mankind (Roy, 1967). This in background, for a country like India, with all its manifestations of socio-economic and technical backwardness has the only alternative way to implement education related programmes on a priority basis. The very realisation resulted in laying emphasis on opening the doors for private sector to spread education at all levels i.e. primary to higher education.

Every individual is born with a collection of abilities and talents. Education in its many forms has the potential to fulfill and help them in order to achieve social and economic development. It has become a common place for national development to be linked to education, with education frequently measured in terms of education (Wagner, 1990).

There is general agreement among scholars that one of the fundamental breakthroughs in the emergence of civilization was the invention of writing as a means of communication. With a written word it has become possible for historical events to be accurately recorded, and for knowledge to be more widely and quickly disseminated among several generations. In writing about importance of education, Goody and Watt (1977) point out that not only was trade, commerce and economic sector of the society radically altered, but also the nature of human interaction transformed.

In brief, the educated person has greater powers of communication, critical consciousness and control over his or her environment. The mobilisation of human potential for social and collective action in ancient Egypt, Babylonia and Greece and even with partially literate populations dramatically surpassed that of pre-literate tribes or nomadic groups. Education also is a basic human right which expands personal choice, control over ones own environment, and allows for collective action not otherwise possible. Much of the concern today in under-developed and developed countries about education stems, at least in part, from this consideration. (Ingemer and Saha,1989).

Changes in the society are coming thick and fast. The changes are explicit in the sectors of economy and technology. They call for new shape of schools, new learner profile, teacher profile, and administrator's profile. More attention has to be given now to excellence, quality and efficiency so that peace and harmony in the society can be maintained. Acceptance and appreciation for diversity and pluralism is inevitable.

The future agenda for education will be to empower individuals, assure high quality of life and pave way to a learning society. Thus, keeping the basic premise of the research investigation under consideration, an attempt was made in this chapter to elucidate on the issues of education, environment issues that play vital role in education and issues related to ethics in one's own life as well as society along with the relevant general ideas that constitute what is management.

a. Higher Education in India: A Perspective

The economic magic of education is widely accepted with developing nations. Education holds miraculous powers in providing upward mobility for the individual and in boosting economic development for a nation. Human Capital theorists (Wagner, 1990) have formalised this faith by offering empirical evidence suggesting that the young 'individual' often does benefit economically from increased levels of schooling within developing nations. Proponents of this theory make the rationalist assumption that the individual (or ones family) maximises his or her economic return in choosing between additional schooling or moving into the labour market. A neo-classical assumption is then made regarding the occupation and wage structure, namely that higher wages are allocated to more productive workers. Therefore, additional investments in human capital formation- by families and nations – lead to higher productivity, private income gains, and aggregate economic expansion (Schultz, 1961).

On the contrary a section of the scholarly view doubts the real contribution of education to economic development. Advocates of alternative proposition suggest that gains in education and schooling follow economic development rather than precede economic growth (Collins, 1977). In addition the most advantaged classes may gain the most from raising levels of education and schooling, reinforcing social and economic inequality (Saltow and Stevens, 1981).

Further, Educationists and Psychologists have focused on discrete elements of reading and writing skills. Such a skill-based approach (termed the 'autonomous model' by Street, 1984) conceptualises education as an autonomous variable whose impact on society and cognition flows from its intrinsic character. This model tends not only to make claims to universalism but privileges the text and the teacher, totally ignoring the knowledge, the learner brings to the learning situation. It doesn't concern itself with the questions of power distribution and authority relations in society nor does it deal with the politics of education.

Whether education contributed to economic development of a nation and how much this contribution to Gross National Product (GNP) - are not particularly useful questions for two reasons. First, it has been evident from what has been learned from experience and studies about education and economic development that the relationship between the two is an interactive one rather than one of the linear causations (Wagner, 1990). The productive model of investigation of physical science applied to the social phenomenon of how education interacts with other aspects of social change does not necessarily enhance knowledge and understanding. Expansion of education or launching of education effort is not independent of other forces of social change. Secondly, societies fortunately do not take decisions on such questions as investment in basic education which impinges on their basic education, values and goals on the basis of rates of return. When such decisions are taken, practitioners in the real world know that the choice is about combinations of physical and human components rather than one or the other (Ibid, 1990).

Education is an important input both for the growth of the society as well as for the individual. Properly planned educational input can contribute to increase in the gross national product, cultural richness, build positive attitude towards technology, increase efficiency and effectiveness of the governance. Education opens new horizons for an individual, provides new aspirations and develops new values. It strengthens competencies and develops commitment. Education generates in an individual a critical outlook on social and political realities and sharpens the ability to self-examination, self-monitoring and self-criticism.

During the last five decades we have gained valuable experiences in all spheres and stages of education in the country. The expectations from education perceived by learners, communities and people are being understood better. The potentialities for future growth are also before us. The expectations and the chance of their being achieved are separated by a wide gap of resource crunch, credibility of institutions, levels of commitment, efficacy of functioning and several other factors.

Contrary to the expectations from education at the time of independence, gaps in education in the context of have and have-nots are increasing. The thin line separating privatisation and commercialisation is getting blurred. Merit alone no longer remains the criteria for moving upwards in education. There is a visible loss of credibility of existing systems of imparting education in schools and also in institutions of higher learning. On one hand we are short of basic infrastructures and on the other, optimum utilisation of existing infrastructures have not been ensured.

Mere appreciation of creating a knowledge society and knowledgeable people is not sufficient. The message must reach each individual that these are the times when every nation needs to move towards a learning society and knowledge society. People in India are better equipped to understand and appreciate it. They are familiar with traditional systems of creating, generating, transferring knowledge and the in-built respect in the society for the learned and the knowledgeable.

Essentially the system of education in a country would always be a social mission with close association between the society and the system of education. A close proximity has always been insisted upon for a viable and indelible system of education to bring in the required positive changes in the society. This in context, a brief attempt has been made in the following section to understand the system of education in India, scope and status of community participation, policies and programmes that are governing the Indian education system and so on have been described. The changes are explicit in the sectors of economy and technology. They call for new shape of educational institutions, new learner profile, teacher profile, and administrator's profile. More attention has to be given now to excellence, quality and efficiency so that peace and harmony in the society can be maintained. Acceptance and appreciation for diversity and pluralism is inevitable.

The future agenda for education will be to empower individuals, assure high quality of life and pave way to learning society. India, like any other knowledge economy, depends on the development of its educational sector. Higher education drives the competitiveness and employment generation in India. However, research findings have shown that the overall state of higher education is dismal in the country. There is a severe constraint on the availability of skilled labour (Agarwal, 2006). There exist socio-economic, cultural, time and geographical barriers for people who wish to pursue higher education (Bhattacharya and Sharma, 2007).

Education is the driving force of economic and social development in any country (Cholin, 2005; Mehta and Kalra, 2006). Considering this, it is necessary to find ways to make education of good quality, accessible and affordable to all, using the latest technology available. The last two decades have witnessed a revolution caused by the rapid development of Information and Communication Technology (ICT). ICT has changed the dynamics of various industries as well as influenced the way people interact and work in the society (UNESCO, 2002; Bhattacharya and Sharma, 2007; Chandra and Patkar, 2007). Internet usage in home and work place has grown exponentially (McGorry, 2002).

India has a billion-plus population and a high proportion of the young and hence it has a large formal education system. The demand for education in developing countries like India has skyrocketed as education is still regarded as an important bridge of social, economic and political mobility (Amutabi and Oketch, 2003). The challenges before the education system in India can be said to be of the following nature:

i. Access to education- There exist infrastructure, socio-economic, linguistic and physical barriers in India for people who wish to access education (Bhattacharya and Sharma, 2007).

ii. Quality of education- This includes infrastructure, teacher and the processes quality.

iii. Resources allocated- Central and State Governments reserve about 3.5% of GDP for education as compared to the 6% that has been aimed (Ministry of Human Resource Development, 2007).

There exist drawbacks in general education in India as well as all over the world like lack of learning materials, teachers, remoteness of education facilities, high dropout rate etc (UNESCO,2002).

Participation of Indian students in education

Stage of Education   Gross Enrolment Ratios (2003-04)
Elementary   85%
Secondary   39%
Tertiary stages of education   9%
(Source: Department of Higher Education, 2007)

Thus, the participation rates of the Indian population in education, and especially in higher education are quite low. In the current Information society, there is an emergence of lifelong learners as the shelf life of knowledge and information decreases. Further, given the population growth and the growing requirements in future, there is a need to provide efficient higher education in India. The efficiency of higher education needs to be both in qualitative and quantitative aspects.

Thus, the higher education system in India needs to be looked into several aspects, especially the ethical and environmental issues which play vital role in this regard. Hence, study of ethical practices in education and the influence of environmental issues, both social and physical, are of much importance in promoting higher education and its management in India.

b. History of Philosophy and Ethics - a brief review:

Philosophy critically examines the conflicting assumptions and upholds the most rational one nearest to truth. This activity and attitude helped to develop knowledge in due course and in this process various theories and branches of knowledge emerged and grown into latest and recent branches of knowledge. Philosophy always aimed at knowing the roots or the fundamentals of anything and everything. The basic search always aimed at knowing the absolute and the ultimate, in its course and quest for knowledge in each and every direction. Ethics is also one of the branches of knowledge that has emanated out of the quest for knowledge in the field of right, wrong, good and bad etc. Clarification of concepts or ideas and meaning and the function of words and phrases is the essential aspect of philosophy. In other words philosophy critically evaluates the assumptions and arguments in depth and also offers clarification of concepts in the process of critical evaluation.

The subject matter of ethics is of two types i.e., in the first place both are beliefs about facts, the first one being what is the case and the second one is about the norms in other words what is to be done. The debate is obviously about an idea of value. Plato (Republic, 352 AD) opined moral philosophy is enquiring "how we ought to live'. By means of studying ethics one will not get the solutions for the problems of life or will aid one as a crutch on which you can lean on but will make one more rational more responsible and more a human being who will be able to contribute for the common good of the society in general.

The story of ethics has got transformed from time to time from the time of Thales(624 BC -546 BC), Anaximander (610 BC- 546 BC, Anaximenes (6th century BC), Demoritus (ca. 460 BC - ca. 370 BC), Lucippus (Greek: first half of 5th century BC), Sophists (5th Century BC and the 2nd Century AD), Socrates(c. 469 BC-399 BC), Plato (428/427 BC - 348/347 BC), Aristotle (384 BC - 322 BC) and to the contemporary thinkers of moral philosophical schools like utilitarianism of John Stuart Mill (1806 -1873), Pragmatism of John Dewey (1859 -1952), Phenomenalism of Jean Paul Sartre(1905 - 1980) and the rest. The common transient streak that runs among all these thinkers is one that is doing the best in the available circumstances in terms of theory and putting it into practice for the welfare and well being of the humanity, during times they have lived. For want of limitations the researcher is only be able to mention the major contributors for the thought of ethics.

The itinerant teachers 'Sophists' when they travelled from city to city in ancient Greece they found that moral code and systems of law differed from place to place and in different societies. This led them to question old beliefs that moral rules are absolute and universal. In this process two concepts are identified one is 'absolute' and the other is 'universal'. When we go through social anthropology, one comes across the fact that the ancient Spartans used to expose weak infants to die on the mountainside. Here we may feel that the custom is cruel from the modern stand point of thinking, but Spartans thought it was right.

c. Ethical values and facts

Value is the concept that majority of the people cherish, desire and aspire which we may call it an 'ideal' which can be called as ethical. It is always on the higher plane than the actual and is always blended with the welfare and well being of the majority and their happiness. The conflict between the ideal and the actual, is the individual personal happiness coming in conflict with the social or common happiness. It's a fact that in the event of a conflict between the ideal and need it is always the ideal that is sacrificed because of the compulsion that the need cannot be postponed. This kind of incapacity on the part of the human beings always lead to a situation that there exist a gap between the ideal and the actual. This is because of the situation that man can not compromise on his personal happiness for the sake of collective happiness in most of the situations. This kind of situation has been prevailing from the day man started his life on earth and will probably last for ever till the end of man's existence.

In the theory of ethics, empiricism appears like that of naturalism. And empiricists rely on experience for true knowledge, but in the case of ethics which is more relevant is not experience or sense perception but that of feeling and desire. In line with this thinking all empiricists of epistemology tend to be naturalists in the theory of ethics. Rationalists treat ethics as statements of contingent truths.

d. Utilitarianism:

Among all schools of philosophy Utilitarianism is one of the most influential school to propagate and contribute for the development of the ethical principles and the subject matter of ethics. In this process utilitarianism believes that an action is right when it is useful in promoting happiness and happiness the theory explains is the sum of pleasures. In other words pleasure is good and displeasure of pain is bad.

An action is considered good and is better when it is capable of generating more pleasure than an action which generates less pleasure. That is nothing but the first act is good when it is considered alone with reference to its capacity to generate pleasure, but in the case of a second act when it generates more pleasure the second is obviously is preferred and is considered good. To sum up utilitarian school of thought believed in, "an action is considered right when it is capable of generating maximum pleasure and is of maximum utility or usefulness to humanity".

Classical utilitarianism (Jermy Bentham of nineteenth century is the exponent) is called as Hedonistic Utilitarianism, which it considered that pleasure alone is good as an end. This form also says that apart from pleasure virtue, knowledge, love and beauty are good as ends and this aspect of thought process is named as Ideal Utilitarianism. Both forms i.e. Classical Utilitarianism and Ideal Utilitarianism are called Utilitarianism because of the only reason that an act to be right is its utility and usefulness and because of their capacity for producing results which are good in themselves. Hedonistic utilitarianism agrees and believes in, that virtue, knowledge, love and beauty are both good by themselves and in themselves and are independent of the goodness of pleasure. Motives like courage, pity, and kindness are considered virtues because of their capacity to generate happiness and are useful for humanity.

Utilitarianism having many attractions has got its own difficulties in the name of intuitionism of rationalists which they proposed as an alternative. Intuitionism can be best being explained in the form that the way to reach the moral judgments is by reason in other words rational intuitionism. Rational intuitionism preaches that

- Promoting happiness of people
- Refraining from harm to people
- Treating people justly
- Speaking truth always
- Keeping promises
- Showing gratitude
- Promoting ones own happiness and
- Maintaining and promoting ones own virtues.

However the greatest good or the fundamental functions of ethics as per Utilitarianism are:

1. Promoting happiness of other people,
2. Treating people justly and
3. Promoting ones own happiness.

e. Ethics of Immanuel Kant

Immanuel Kant (1724 -1804) was born in Konisberg, Germany. In spite of its various attractions Utilitarianism is able to give only a reasonably accurate picture of everyday moral judgements and unable to meet the needs of philosophical theory and as a coherent system of ethics in a comprehensive way.

This formulation was stated by Kant who distinguished the categorical imperative (moral) from Hypothetical (prudential or technical) Imperative. A hypothetical imperative has the form of 'Do X or ought to do X if ... for example 'read books if you want to acquire knowledge'. The categorical imperative does not depend on 'If' and a means to an end by itself. Kant has given three types of formulations i.e. the fundamental principle of moral action (categorical imperative) and some times he wrote as if there were four formulations but two of these were on a single theme and were variations and to have a threefold formula shows a clear aspect of the theory. The first is concerned with the form from the categorical imperative the second is concerned with the concept and the third links these together.

1. Act as if you were legislating for everyone.
2. Act so as to treat human beings always as ends and never merely as means. and
3. Act as if you were a member of a realm of ends.

Kant's categorical imperative concerns with the form and gives rise to the statements of universal moral judgements like 'ought to do' in nature. That means anyone and every one, in certain set of conditions has to take the same course of action or to do the same in the given circumstances which acquires universality in nature.

In the second formulation that is the hypothetical imperative Kant substantiates or provides reason for 'right action' and treats man as an end but never as a means. To treat man as an end is to make his ends your own and regard his choices as your own and deal accordingly. To sum up this kind of attitude on the rationalization of ethical and moral principles on the part of Kant appears to be similar to the biblical commandment 'love thy neighbour as thyself'.

In the third formulation Kant connects the first and second formulations that are categorical imperative and the hypothetical imperative. And accordingly we can sum up that the Kantian ethics wants one to feel like a man of this society who got the similar desires, destinies and dispositions in this world. This kind of treatment of ethical principles, by Kant has given rise to ethical concepts like democracy, justice, liberty, fraternity very strongly and subsequently called as an ethics of democracy.

f. Main trends of ethics in 20th century:

In general, in general and cultural life various doctrines that evolved in the west are by and large are pluralistic in nature, and they addressed both theoretical and practical aspects of ethical problems. Given the resources and constraints at the disposal of the investigator it is beyond the capacity to indulge in a detailed chronological depiction of the process of the ethical trends of 20th century. Accordingly the investigator attempted modestly to give a brief account of the dominant schools of ethics of contemporary times. In this process the review is confined to ethics of, irrationalism, formalism, and naturalism and axiological and theological schools rather than the subject matter of these schools.

g. **Existentialist Ethics**

Existentialism is a widespread and one of the dominant philosophical trends. In Latin 'existentia' means existence. This school of thought has emanated predominantly after the First World War which has taken up the continued philosophical trend of 19th century. This school of thought was represented by Nietzsche (1844) and Kierkegaard (1813 -1855). It mostly conveys that of irrationalism in western social thought. The towering personalities who contributed for this school are Karal Jaspers (1883 - 1969), Martin Heidegger (1889 -1976), Gabriel Marcel (1889-197), Jean Paul Sartre, and Albert Camus (1913 -1960). Existentialism unlike other philosophical schools places a special emphasis on morality. The entire philosophical literature is moral in nature. Schopenhauer recognized that the world was determined by the subject and also at the same time attached significance to will that exceeded the boundaries of human existence and assumed a cosmic meaning. Moral imperative is given an ontological status by existentialism. Morality as interpreted by existentialism that social morality is unreal and 'real' morality lies outside the society.

h. Pragmatism

Charles pierce (1839-1914) is the first person to use the word 'Pragmatism' and this school of thought was developed mostly in1800 AD in United States of America and Peirce developed the idea that inquiry depends on real doubt, not mere verbal or hyperbolic doubt and said, in order to understand a conception in a fruitful way, "Consider what effects that might conceivably have practical bearings you conceive the objects of your conception to have. Then, your conception of those effects is the whole of your conception of the object - which he later called the pragmatic Maxim". In other words the goodness or the badness of an action would be determined by means of its consequences, and accordingly be called pragmatic or not.

According to pragmatism, theory and practice are not separate spheres. John Dewey has put it that there is no question of theory versus practice but good practice versus bad practice. Pragmatist ethics is broadly humanistic because it sees no ultimate test of morality beyond what matters for us as humans. In his classic article 'Three Independent Factors in Morals' John Dewey (1930), tried to integrate three basic philosophical perspectives on morality: the right, the virtuous and the good. He held that while all three provide meaningful ways to think about moral questions, the possibility of conflict among the three elements cannot always be easily solved. John Dewey stressed the need for meaningful labour and a conception of education that viewed it not as a preparation for life but as life itself.

Pragmatism treats that there is no difference between facts and values. Both facts and values have cognitive content: knowledge is what we should believe; values are hypotheses about what is good in action. Pragmatist ethics is broadly humanist because it sees no ultimate test of morality beyond what matters for us as human beings. Good values are those for which we have good reasons.

Dewey did not support the dichotomy between means and ends which he saw as responsible for the degradation of our everyday working lives and education, both conceived as merely a means to an end. He stressed the need for meaningful labour and a conception of education that viewed it not as a preparation for life but as life itself. [Dewey1938- 1997]

i. Ethics: A Perspective

As ethical practices too play vital role in promoting and strengthening higher education, an attempt was made to understand the basic issues related to Ethics in the following manner.

Many people tend to equate ethics with their feelings. However, being ethical is clearly not a matter of following one's feelings. A person following his or her feelings may recoil from doing what is right. In fact, feelings frequently deviate from what is ethical.

Nor should one identify ethics with religion. Most religions, of course, advocate high ethical standards. Yet if ethics were confined to religion, then ethics would apply only to religious people. But ethics applies as much to the behaviour of the atheist as to that of the saint. Religion can set high ethical standards and can provide intense motivations for ethical behaviour. Ethics, however, cannot be confined to religion nor is it the same as religion.

Being ethical is also not the same as following the law. The law of US often incorporates ethical standards to which most citizens subscribe. But laws, like feelings, can deviate from what is ethical. Our own pre-Civil War slavery laws and the apartheid laws of present-day South Africa are grotesquely obvious examples of laws that deviate from what is ethical.

Finally, being ethical is not the same as doing "whatever society accepts." In any society, most people accept standards that are, in fact, ethical. But standards of behaviour in society can deviate from what is ethical. Moreover, if being ethical were doing "whatever society accepts," then to find out what is ethical, one would have to find out what society accepts. To decide what one should think about abortion, for example, one would have to take a survey of American society and then conform one's beliefs to whatever society accepts. But no one ever tries to decide an ethical issue by doing a survey. Further, the lack of social consensus on many issues makes it impossible to equate ethics with whatever society accepts. Some people accept abortion but many others do not. If being ethical were doing whatever society accepts, one would have to find an agreement on issues which does not, in fact, exist.

What, then, is ethics? Ethics is two things. First, ethics refers to well based standards of right and wrong that prescribe what humans ought to do, usually in terms of rights, obligations, benefits to society, fairness, or specific virtues. Ethics, for example, refers to those standards that impose the reasonable obligations to refrain from rape, stealing, murder, assault, slander, and fraud. Ethical standards also include those that enjoin virtues of honesty, compassion, and loyalty. And, ethical standards include standards relating to rights, such as the right to life, the right to freedom from injury, and the right to privacy. Such standards are adequate standards of ethics because they are supported by consistent and well founded reasons.

Secondly, ethics refers to the study and development of one's ethical standards. As mentioned above, feelings, laws, and social norms can deviate from what is ethical. So it is necessary to constantly examine one's standards to ensure that they are reasonable and well-founded. Ethics also means, then, the continuous effort of studying our own moral beliefs and our moral conduct, and striving to ensure that we, and the institutions we help to shape, live up to standards that are reasonable and solidly-based.

Ethics is concerned with what is right or wrong, good or bad, fair or unfair, responsible or irresponsible, obligatory or permissible, praiseworthy or blameworthy. It is associated with guilt, shame, indignation, resentment, empathy, compassion, and care. It is interested in character as well as conduct. It addresses matters of public policy as well as more personal matters. On the one hand, it draws strength from our social environment, established practices, law, religion, and individual conscience. On the other hand, it critically assesses each of these sources of strength. So, ethics is complex and often perplexing and controversial. It defies concise, clear definition. Yet, it is something with which all of us, including young children, have a working familiarity.

This makes ethics sound like morality. This is intentional on our part. Like most contemporary texts, ours will treat ethics and morality as roughly synonymous. This is in keeping with the etymology of the two words. Moral derives from the Latin word 'moralis'. Moralis was a term that ancient Roman philosopher Cicero made up to translate the ancient Greek 'Ethikos' into Latin. Both mean, roughly, pertaining to character; but today their English derivatives deal with much more than character.

It is tempting to seek a general definition of ethics before discussing any particular ethical topic. Although it has been said a little bit about what we take ethics to be, we have not offered such a definition; and we will not do so. Demanding a definition at the outset can stifle discussion as easily as it can stimulate it. We offer one of Plato's dialogues as a case in point.

In the Euthyphro we find Socrates and Euthyphro meeting each other on the way to court. Socrates is being tried allegedly for corrupting the youth by encouraging them to believe in "false gods" (Dimon) and for making the better argument appear the worse. Euthyphro is setting out to prosecute his father allegedly for murdering one of his servants. Socrates expresses surprise that Euthyphro would prosecute his own father, and he asks him for an explanation. Euthyphro appeals to the justice of doing this. Socrates then asks him to define justice. Euthyphro offers some examples of justice and injustice. Socrates rejects them all on the grounds that they are only examples, whereas what he wants Euthyphro to tell him is what all just acts have in common that makes them just. That is, what Socrates demands is a definition that captures the essence of justice in all of its instances. Unfortunately, Euthyphro attempts to satisfy Socrates' demand rather than challenge its reasonableness. All of his efforts fail miserably, and the dialogue ends with Euthyphro indicating he must leave to get on with his business. The implication is that Euthyphro is going off to prosecute his father without the least grasp of the value in which name he is acting, justice.

As much as we might desire the sort of definition Socrates and Euthyphro were seeking, it seems an unreasonable demand. At best, this might come at the end of an inquiry rather than at its beginning. Morality, like science, should allow room for piecemeal exploration and discovery. It should not be necessary to provide a comprehensive definition of justice in order to be able to say with confidence that sometimes drawing lots is a just procedure, having the person who cuts the pie get the last piece is just, compensating people for the work they do is just, denying women the right to vote is unjust, punishing the innocent is unjust, and so on. Further reflection might reveal special features these examples all have in common, or at least special ways of grouping them. But having a solid starting point does not require having a well worked out definition of the concept under consideration.

18th century philosopher Thomas Reid has some useful piece of advice for those interested in developing a systematic understanding of morality. He compares a system of morals to "laws of motion in the natural world, which, though few and simple, serve to regulate an infinite variety of operations throughout the universe." However, he contrasts a system of morals with a system of geometry: A system of morals is not like a system of geometry, where the subsequent parts derive their evidence from the preceding, and one chain of reasoning is carried on from the beginning; so that, if the arrangement is changed, the chain is broken, and the evidence is lost. It resembles more a system of botany, or mineralogy, where the subsequent parts depend not for their evidence upon the preceding, and the arrangement is made to facilitate apprehension and memory, and not to give evidence.

Reid's view has important implications for how we should characterize moral development. On the botanical model, access to basic moral understanding need not be an all or nothing affair. Its range and complexity can be a matter of degree, and confusion in one area need not infect all others. Understanding how different, basic moral considerations are related to one another can be a matter for discovery (and dispute) without our having to say that those whose picture is incomplete or somewhat confused have no understanding of basic moral concepts.

j.   Ethics and Childhood

Children's introduction to ethics, or morality, comes rather early. They argue with siblings and playmates about what is fair or unfair. The praise and blame they receive from parents, teachers, and others encourage them to believe that they are capable of some degree of responsible behaviour. They are both recipients and dispensers of resentment, indignation, and other morally reactive attitudes. There is also strong evidence that children, even as young as four, seem to have an intuitive understanding of the difference between what is merely conventional (e.g., wearing certain clothes to school) and what is morally important (e.g., not throwing paint in another child's face). So, despite their limited experience, children typically have a fair degree of moral sophistication by the time they enter school.

What comes next is a gradual enlargement and refinement of basic moral concepts, a process that, nevertheless, preserves many of the central features of those concepts. All of us can probably recall examples from our childhood of clear instances of fairness, unfairness, honesty, dishonesty, courage, and cowardice that have retained their grip on us as paradigms, or clear cut illustrations, of basic moral ideas. As philosopher Gareth Matthews puts it: A young child is able to latch on to the moral kind, bravery, or lying, by grasping central paradigms of that kind, paradigms that even the most mature and sophisticated moral agents still count as paradigmatic. Moral development is ... enlarging the stock of paradigms for each moral kind; developing better and better definitions of whatever it is these paradigms exemplify; appreciating better the relation between straightforward instances of the kind and close relatives; and learning to adjudicate conflicting claims from different moral kinds (classically the sometimes competing claims of justice and compassion, but many other conflicts are possible).

This makes it clear that, although a child's moral start may be early and impressive, there is much conflict and confusion that needs to be sorted through. It means that there is a continual need for moral reflection, and this does not stop with adulthood, which merely adds new dimensions.

Nevertheless, some may think that morality is more a matter of subjective feelings than careful reflection. However, research by developmental psychologists such as Jean Piaget (1896-1980), Lawrence Kohlberg (1927-1987), Carol Gilligan (1936), James Rest (1932-1999), and many others provides strong evidence that, it is important as feelings are; moral reasoning is a fundamental part of morality as well. Piaget and Kohlberg, in particular, did pioneering work to show that there are significant parallels between the cognitive development of children and their moral development. Many of the details of their accounts have been hotly disputed, but a salient feature that survives is that moral judgment involves more than just feelings. Moral judgments (e.g., "Smith acted wrongly in fabricating the lab data") are amenable to being either supported or criticized by good reasons. ("By fabricating the data, Smith has misled other researchers and contributed to an atmosphere of distrust in the lab." "A thorough examination of Smith's notebooks shows that no fabrication has taken place.")

Kohlberg's (1927-87) account of moral development has attracted a very large following among educators, as well as a growing number of critics. He characterizes development in terms of an invariable sequence of six stages. The first two stages are highly self-interested and self-centred. Stage one is dominated by the fear of punishment and the promise of reward. Stage two is based on reciprocal agreements ("You scratch my back, and I'll scratch yours"). The next two stages are what Kohlberg calls conventional morality. Stage three rests on the approval and disapproval of friends and peers. Stage four appeals to "law and order" as necessary for social cohesion and order. Only the last two stages embrace what Kohlberg calls critical, or post-conventional, morality. In these two stages one acts on self-chosen principles that can be used to evaluate the appropriateness of responses in the first four stages. Kohlberg has been criticized for holding that moral development proceeds in a rigidly sequential manner (no stage can be skipped, and there is no regression to earlier stages); for assuming that later stages are more adequate morally than earlier ones; for being male biased in overemphasizing the separateness of individuals, justice, rights, duties, and abstract principles at the expense of equally important notions of interdependence, care, and responsibility; for claiming that moral development follows basically the same patterns in all societies; for underestimating the moral abilities of younger children; and for underestimating the extent to which adults employ critical moral reasoning. It is not attempt to address these issues here.

Nevertheless, whatever its limitations, Kohlberg's theory makes some important contributions to our understanding of moral education. By describing many common types of moral reasoning, it invites us to be more reflective about how we and those around us typically do arrive at our moral judgments. It invites us to raise critical questions about how we should arrive at those judgments. It encourages us to be more autonomous, or critical, in our moral thinking rather than simply letting others set our moral values for us and allowing ourselves to accept without any questions the conventions that currently prevail. It brings vividly to mind our self-interested and egocentric tendencies and urges us to employ more perceptive and consistent habits of moral thinking. Finally, it emphasizes the importance of giving reasons in support of our judgments

k. Descriptive and Normative Inquiry

It is useful to think of ethics, or morality, as an umbrella term that covers a broad range of practical concerns, many of which are rather straightforwardly understood and dealt with, but some of which are not very clearly understood and are often quite controversial. This can help us see how the study of ethics differs from most other subjects of study, at least as they are traditionally understood.

Chemistry, for example, is typically viewed as empirical, or descriptive. We study chemistry to learn about how acids are different from bases, what the basic chemical properties of certain metals are, what the most basic principles are that explain chemical changes, and so on. Presumably, what we learn is based on careful, scientific observation. There is an attempt to describe what the case is, at least in the world of chemistry.

There is a descriptive aspect of morality too. Psychologists, sociologists, and anthropologists might try to determine which particular values a certain group of people actually accept, how these values are related to people's behaviour, their social and political institutions, or their religious beliefs. They can assemble information about the kinds of values people hold. Some of these values, although not moral values by themselves (e.g., certain aesthetic values or value we attach to material goods), may nevertheless be regarded as important enough to be accorded moral (and even legal) protection. But social scientists can describe this without necessarily endorsing the values that people actually accept as values they ought to accept. To ask what values people ought to accept is to ask a normative, rather than simply a descriptive question. It is to ask what values are worthy of being accepted, rather than simply whether they are accepted; and it is the business of normative ethics to address these questions.

1. Philosophical Ethics

Traditionally, ethics has been taught at the college level mainly in departments of philosophy. In large part, philosophical ethics is normative in its focus. It examines basic questions about what our values should be, what, if any, fundamental grounding they can be given, and whether they can be organized into a comprehensive, coherent theory. Another part of philosophical ethics is called meta-ethics, which studies the nature of the language and logic we use when we are concerned about morality (as distinct from, say, law or social etiquette).

Although the study of philosophical ethics might make valuable contributions to our understanding of relationships between ethics and science, we do not regard it as a necessary preparation for bringing ethics into science classes. Thomas Reid wisely warns us not to make the mistake of thinking that in order to understand [one's] duty; [one] must or need be a philosopher and a metaphysician. This does not mean that careful reflection is not needed. Nor does it mean that philosophical reflection is not needed. But, just as we do not need to be logicians in order to think logically, mathematicians in order to think mathematically, or scientists in order to think scientifically, we do not have to be philosophers in order to think philosophically.

What Reid is telling us is that we do not need to be a Plato or Aristotle in order to know our way about morality. He is also telling philosophers that in framing their theories they need to respect the understanding that ordinary, thoughtful people have of morality even though they may never have opened a philosophy book. In fact, most moral philosophers do this. For example, Aristotle's account of the virtues, Immanuel Kant's categorical imperative ("Act only on those maxims that you could at the same time will to be a universal law"), and John Stuart Mill's utilitarian theories (promote the greatest good for the greatest number) all begin with what they take to be commonly accepted moral views; and they see their task as articulating, refining, and reworking these views where necessary. They do this in ways that, nevertheless, respect common, everyday morality. For example, Kant tries to show how his categorical imperative gives us an improved understanding of the moral insights provided by the Golden Rule. John Stuart Mill argues that his utilitarian theory both respects and provides a solid foundation for such basic, commonly accepted rules of morality as telling the truth and keeping promises, while at the same time providing a more fundamental principle for resolving conflicts among rules (e.g., when keeping a promise requires harming someone). Difficult to discern as their writings sometimes are, the constraints that common morality placed on them remain evident.

m. Common Moral Values

Given the apparent moral differences found among people with different national, ethnic, or religious backgrounds, it may seem naive to talk, as we have, of common moral values. What moral values, if any, might be sharable across national, ethnic, religious, or other boundaries? This is the question philosopher Ms. Sissela Bok (1934) takes up in her recent book, Common Values. She begins by listing a number of problems that cut across these boundaries: problems of the environment; war and hostility; epidemics; overpopulation; poverty; hunger; natural disasters (earthquakes, tornados, drought, floods); and even technological disasters (Chernobyl). The fact that we recognize these as common problems suggests that we share some basic values (e.g., health, safety, and the desire for at least minimal happiness).

However, our desire to get to the bottom of things often blocks gaining a clearer understanding of what we have in common. Ms. Bok nicely outlines this problem. She notes that we may feel we need a common base from which to proceed. But there are different ways in which we might express what we think we need. Ms. Bok mentions ten different ways. We may seek a set of moral values that are

1. Divinely ordained
2. Part of the natural order
3. Eternally valid
4. Valid without exception
5. Directly knowable by anyone who is rational

6. Perceivable by a "moral sense,"

7. Independent of us, in the sense that they do not depend on us for their existence.

8. Objective rather than subjective

9. Held in common by virtually all human beings

10. Such that they've had to be worked out by all human societies.

Although religious and philosophical traditions have concentrated on points from 1 to 8 above, Ms. Bok suggests we should start with point 9 and 10. Given the inability of our religious and philosophical traditions to reach consensus thus far on points from 1 to 8, it seems unlikely, she says, that we will reach consensus on points from 1 to 8.

In regard to 9 and 10, Ms. Bok makes four basic claims. First, there is a minimalist set of values that every viable society has had to accept in order to survive collectively. This includes positive duties of mutual support, loyalty, and reciprocity; negative duties to refrain from harming others; and norms for basic procedures and standards for resolving issues of justice. Second, she says that these values are necessary (although not sufficient) for human coexistence at every level-in one's personal and working life, in one's family, community, and nation, and even in international relations. Third, these values can respect diversity while at the same time providing a general framework within which abuses can be criticized. Finally, Bok says, these values can provide a common basis for cross-cultural discussions about how to deal with problems that have global dimensions.

Ms. Bok's point about finding common values while respecting diversity is very important. It is fairly easy to see that the same general values might play themselves out quite differently from one locale to another. For example, although England and the United States drive on opposite sides of the road, they share the same basic values of safe and efficient travel. There is no reason to insist that one way is better than the other for these purposes. However, either is clearly preferable to, say, a rule that mandates driving on the left side on Monday, Wednesday, Friday and the right side on Tuesday, Thursday, and the weekend -- or no rule at all. The United States tends to use stop lights at intersections, while England favours roundabouts. They may work equally well, or one may be better than the other -- as judged by the same general values of safety and efficiency. It is also quite likely that both systems can be improved in ways yet to be discovered.

However, Ms. Bok is making another point as well. She is suggesting that, even in the absence of agreement at the most fundamental level, those with very different moral and religious backgrounds may find common ground. A good example of this is the consensus reached by the National Commission for the Protection of Human Subjects of Biomedical and Behavioural Research. This commission was established by the United States Congress in 1974, and it issued what is known as the Belmont Report in 1978. This report contains the guidelines used by Institutional Review Boards (IRB's) at colleges, universities, and other institutions that receive federal funding for research involving human subjects. The task of IRBs' is to examine research protocols to make certain that the rights and welfare of human subjects are being protected.

Congress made a serious effort to ensure that different perspectives would be represented. Albert Jonsen and Stephen Toulmin describe the composition of the commission in this way: The eleven commissioners had varied backgrounds and interests. They included men and women; blacks and whites; Catholics and Protestants, Jews, and atheists; medical scientists and behaviourist psychologists; philosophers; lawyers; theologians; and public representatives. In all, five commissioners had scientific interests and six did not.

The commission got off to a slow start. Their deep religious and philosophical differences surfaced quickly and blocked their ability to move ahead. Then they decided to talk first about specific examples rather than more foundational concerns. As they discussed particular cases of research involving human subjects (like the Tuskegee case), they discovered substantial areas of agreement that enabled them eventually to formulate three basic areas of ethical concern: respect for persons, beneficence, and justice.

In articulating their concerns about respect for persons, the commission agreed with the Kantian idea that it is inappropriate to treat persons merely as means to the ends of research. They agreed that it is important to obtain the informed consent of subjects before including them in an experiment, thus respecting their ability and right to make an informed decision (respect for autonomy). In regard to beneficence, the commission accepted the utilitarian idea of trying to maximize benefits to human subjects while minimizing the risk of harm. Finally, in regard to justice, the commission agreed that discrimination in the selection of research subjects is inappropriate and that special attention needs to be given to especially vulnerable groups such as prisoners, children, and the elderly.

However, the commission also carefully avoided committing itself to a set of inflexible guidelines. The Belmont Report confidently, but modestly, comments: Three principles, or general prescriptive judgments, that are relevant to research involving human subjects are identified in this statement. Other principles may also be relevant. These three are comprehensive, however, and are stated at a level of generalization that should assist scientists, subjects, reviewers and interested citizens to understand the ethical issues inherent in research involving human subjects. These principles cannot always be applied so as to resolve beyond dispute particular ethical problems. The objective is to provide an analytical framework that will guide the resolution of ethical problems arising from research involving human subjects.

So, as a result of their willingness to reason with each other despite their differences, the commission succeeded in coming up with a workable document that is now reflected in the policies and practices of research institutions that receive federal funding for some of their research. Both the deliberate process and its results bear the marks of reasonableness that we might hope is obtainable in a democratic, but diverse, society. In fact, the work of the commission models many of the values that can be served by bringing ethics into the science classroom by making apparent how science and ethics are interrelated and how the challenges this poses might be thoughtfully addressed.

n. Environment for Developing Higher Education

To sum up, ethics has omnipresence in human life and so do its influence on higher education. Though ethical practices have its impact on Higher education in India yet there are other factors which also assert impact on the same. In fact, these factors are nothing else but a conglomeration of various issues drawn from social, economic, ethics, values, traditions, culture and other related aspects.

In fact, prior to independence, the growth of institutions of higher education in India was very slow and diversification in areas of studies was very limited. After independence, the number of institutions has increased significantly. There are today, two hundred and fourteen (214) universities and equivalent institutions including one hundred and sixteen (116) general universities, twelve (12) science and technology universities, seven (7) open universities, thirty three (33) agricultural universities, five (5) women's universities, eleven (11) language universities, and eleven (11) medical universities. Besides, there are universities focusing on journalism, law, fine arts, social work, planning and architecture and other specialized studies. In addition, there are 9703 colleges where 80% of undergraduate and 50% of postgraduate education is imparted. The number of students has reached the level of 6.75

million and there are 3, 21,000 teachers in the higher education system. The government expenditure alone was of the order of Rs. 42,126 millions in 1996-97, and during the subsequent period this has risen even higher.

A special emphasis has come to be laid on women's education. The number of women's colleges has recorded a substantial increase, and India has 1195 women's colleges today. The enrolment of women at the beginning of 1997-98 was 2.303 million, 34 per cent of them being of the postgraduate level.

This massive development has been guided by a process of planning and recommendations of several national commissions set up by the Government of India. The objectives of higher education have gradually become more and more precise and a system of governance is developing in the direction of increasing autonomy and accountability.

But, in spite of vast efforts over the last 50 years, it is only now that the country is slowly emerging out of the fetters of old ideas and rigid structures, built during the colonial rule. There is at present a demand for radical changes which have the potential to actualise a national system of education that was visualised during the freedom struggle.

Learning is a complex process and it involves not only children and their teachers but families as well. Factors affecting the achievements of pupils can be broadly categorized into school-related and household related. Generally, children who are fortunate in being born to educated parents or having caring, competent teachers do very well, and are able to find jobs demanding high productivity. However, the average is appallingly low. The results are low productivity, poor skills, and massive unemployment even after several years of schooling, or even college education.

Various studies have reported that children coming from a deprived background do not have a supportive learning environment and feel alienated in schools. The government school teachers, even motivated ones, find it difficult to address their special needs. Therefore, increasingly it is being realized that only by improving the quality of education can the positive effects of growing enrolments be sustained.

The Public Report on Basic Education (PROBE) which investigated the schooling situation in over 200 villages of north India in 1996, says, "quality education", however defined, involves certain minimal requirements such as adequate facilities, responsible teachers, an active class room and an engaging curriculum. These are simply not met at present.

India, like other countries, visualizes that a new age is dawning, that will be characterised by unimaginable advances in knowledge and synthesis of knowledge, triggering major changes in the objectives, contents, and methods of higher education. Great emphasis will fall upon lifelong education and the realization of a learning society. Complete education for the complete human personality will come to be emphasized more and more imperatively. Building up the defences of peace in the minds of men and women will continue to make tremendous demands on all levels of education, and higher education will have to bear the responsibility.

India also visualizes that contemporary problems can be resolved only if human nature is so changed that mutual goodwill and spontaneous drive to cooperation become ingrained in the human consciousness. India, therefore, visualizes a number of tasks that relate to the creation of a new society that is non-exploitative and non-violent in character by virtue of the integrated personalities of the constituents.

The educational environment which influences the Higher education can be considered in a three dimensional way. The first dimension is the student itself. How the student conceives the teaching methods, teachers, attitude of parents etc. is one dimension which influences spread of education. Further, since Higher education is a litmus test for the student to acquire basic skills for earning for himself during the rest of his life, he considers a whole lot of issues which are mostly dominated by psychological, physical and management issues that prevail with the student. The other dimension is the issue of the social and physical atmosphere prevailing in the education institution wherein the student is pursuing his interests in Higher education. The way teachers adopt different methods of teaching, their approach towards the students and lot of other issues that emanate from the very educational institution. The third dimension is the parents of the students itself. It is the influence of parents on the student matters most. Their approach to the student in terms of psychological and social aspects, economic supports they provide etc. do play vital role in pursuing higher education by the student.

The recent modern new tide of educational thought which aims at drawing from our current experience of all that is quintessential, as also to develop a new vision in the light of the highest traditions of Indian education and of the contemporary needs and aspirations.

The following educational objectives are being emphasised:

- Education aims at liberation — liberation from bondage and ignorance, backwardness and gravitational pulls of the lower human nature;

- Education, being an evolutionary force that enables both the individual and the collectivity to evolve various faculties and to integrate them by the superior intellectual, ethical, aesthetic and spiritual powers, should aim at developing a new type of humanity highly humane, cultured and integrated.

- Education should be developed as a harmonising force, which tries to relate the individual, environment and cosmos in a total harmony by the purification and cultivation of various domains of outer space and inner space;

- Education should be so designed as to become a powerful carrier of the best of the heritage and it should, therefore, aim at transmitting to the new generations the lessons of the accumulated experiences of the past for further progress in the present and the future.

Considering that the contemporary problems of environment, of conflicts and of asymmetrical relationships need to be resolved as early as possible, the Indian system of education aims at the promotion of the goals of universal peace, harmony and unity, based on the principles of liberty, equality and fraternity.

Thus, the process of education, especially Higher education is a matter of complex issue wherein certain ethical practices, parental influence and the education institution itself play a concurrent role in promoting the same. Hence, an attempt was made through this research study to understand the ethical practices being followed by the educational institutions and their management as well as students while pursuing higher education.

o. **Management a perspective**

Management of higher education in India is being viewed as a powerful tool for fostering the development in the recent years by the rulers and administrators. Management also plays a vital role in judiciously administering the meagre and valuable resources and help obtain the desired results in achieving the nation's goals at macro and micro levels. In order to understand the basic tenets of Management, a brief review of Management Thought and its transformation over years is attempted in nutshell.

p. **Evolution of Management:**

The concept of management and its practice can be traced back to 3000 B.C to the first government organization developed by the Sumerians and Egyptians. The early study of management as we call it today is 'Classical Perspective' of Management. In the following few pages a brief review of management and its development and thought is attempted to by the researcher.

Management emphasizes on a constructive and systematic approach and methodology for any organisation. The management of educational institutions require some principles to be adopted in shaping the students to become good citizens. In education system besides teaching the student should be inculcated moral values, along with basic management tenets should be embedded in the system. In line with this the ideas of some of the management thinkers are presented in the following lines.

1. Frederick Winslow Taylor (1856 -1915)

Classical perspective of Management has stressed the need for rational and scientific approach to study and problem solving that emerged during the late nineteenth (19) and early twentieth (20) century which sought to make organizations efficient machines. With the beginning of the factory systems in 1800s the approach to pinpoint organizational plans, tools, structures, systems, jobs, employee's roles and strict adherence to objectives/goals achievement became the basic pre requisite. The various aspects of modern management is predominently oriented towards production and production oriented factory setups which started mostly in the west during the early 19th century and picked up momentum after the second world war. In this regard Frederick Winslow Taylor (1856 -1915) through his theory regarding improvement of productivity on labour has earned him the status of 'Father of Scientific Management'. He preached for, deviating from thumb rules to precise measurement of procedures for various aspects of work and related activities from the stage of planning and its various stages of transformation into product and beyond marketing and service as well. In this process Taylor's statement that 'in the past man has been the first but in the future the system must be first' was acclaimed by one and all and reflects his thinking on the need for precision in assessment (Planning) as well as up to the stage of execution.

2. Henry Gantt L (1893) - Scientific Management:

It is a sub field of classical management perspective and emphasized and strove hard to scientifically determine changes in management practices as the solution to improving labour productivity. Henry Gantt an associate of Taylor has developed one bar graph that measures planned and actual/completed work along each stage of production by time frame or time elapsed.

Frank B Gilbreth & Lillian M Gilbreth (1868 -1924) both wife and husband', the former being a mason and his wife being an industrial psychologist pioneered time and motion study, i.e., the former has divided work into various movements and the later provided the scientific measurement with reference to time and level of effort in case of different work situations.

3. Max Weber (1864 - 1920):

Max Weber a German theorist introduced the most of the concepts on bureaucracy as the systematic approach to management started developing in the Europe during the late 1800s. As most of the European businesses are developed as family businesses the orientation and loyalty of the individuals focused mostly to individuals rather than that of the organizations. In this regard it is worth noting that there existed a hierarchy in the organizations and the individuals and individual goals attained more importance than the group or organizational goals.

4.   Mary Parker Follett (1868 - 1933):

She was a major contributor for the administrative principles' approach to Management. Her ideas were followed by the businessmen because of their practical utility and result orientedness which were relatively overlooked by management scholars. Her emphasis on worker participation and shared goals among managers was followed by many contemporary business people of the time which was recently discovered by Corporations of USA.

The most significant work in this direction 'General and Industrial Management' of Henry Fayol (1841 - 1925) discussed fourteen (14) general principles of management several of them happen to be the regular practices of the modern management philosophy even today.

Unity of command - is the concept that each worker/employee has to receive instructions/orders only from one - superior or boss.

Division of work - is a work technique that workers are amenable and are capable of producing more with the same amount of work/effort because of specialization.

Unity of Direction – same type of activities in an organization are to be grouped together **in an organization under one manager.**

Scalar Chain – a chain of authority that starts from the top of the organization and ends at the bottom and includes each and every employee irrespective of his trade, education, skill, or designation.

5.   Chester I Barnard (1868 – 1961):

He has contributed two important concepts in management i.e., informal organization and the acceptance theory of authority which state that people are not mere machines and people follow orders as they perceive positive benefits by doing so. And this concept of informal organization plays a crucial role in organizational success. In general classical perspective of management has contributed successfully for the productivity and industrial growth and development of United States of America and other countries especially Japan borrowed from USA and got benefited as well.

6.   Behavioural Sciences Approach-Humanistic Perspective – Human relations movement:

Behavioural Science Approach of management has come into existence in the organizational context by drawing heavily from the principles of the social sciences viz. Economics, Psychology, Sociology and other subjects. One specific set of management technique that is predominantly based on behavioural sciences is the concept of 'Organization Development' during 1970s and since, been broadened and expanded with the increasing complexity of organizations and the environments both internal and external. In this regard Hawthorne studies (1924) a series of experiments were started to establish the cause and effect relationship with reference to various variables related to productivity of workers, in Hawthorne plant of Western Electric Company, Illinois USA. They started with that better treatment of the experimental group results in better productivity, but to their surprise the variables other than the controlled variables like human relations, team spirit etc. but not money proved to be the motivators. This they coined it 'Hawthorne effect' and a movement of management thinking and practice that emphasized, satisfying employee needs is the key to improved productivity. This humanistic approach for management vouched for designing the jobs in such a way to bring out the best from the workers and to utilize their full potential. The advocates of this movement were Abraham Maslow (1908 -1970) who proposed his hierarchy of needs theory from 'physiological needs to self actualization being the highest need'.

7.   Douglas McGregor (1906 -1964):

He proposed the famous X theory and Y theory, wherein X theory views workers with negative assumptions that workers detest work and to be forced to work and to be closely supervised, etc. and Y theory with positive assumptions that 'workers like work and enjoy it and keep on producing without any supervision with self motivation'.

8.   Management Science Perspective:

Management Science Perspective emerged after the Second World War with a strong reliance on the subjects like applied mathematics, statistics, and other quantitative techniques being resorted to, for finding solutions for the management problems in organizations. The spirit of management, revolved around productivity, effectiveness and efficiency. The dominant fields of management that emanated are, operations research, operations management, and information technology and the connected sub branches of knowledge like mathematical models, forecasting methods, linier and non-linier programming, queuing theory, scheduling, simulation, Enterprise Resource Planning, e-commerce etc.

Management theory by employing all the above methods strove hard to find solutions for anything and everything that is connected with production of goods as well as services. The range of activity for this approach includes both production of goods and services that are connected with human being and its existence. In this regard education is also in no way different from the innumerable services that falls under the scope of management.

9. Recent Trends in Management:

Management by virtue of its nature and content is dynamic and seeks finding solutions for the problems that arise out of each and every situation. In other words management always preaches 'The Only Right Way Always' in approach of the problem and finding the solutions by utilizing the optimum of the resources and getting always the best output in any given situation, be it quality, quantity, time or resources. With population explosion and its ever growing needs have given an impetus for this management approach of 'optimization of resources and maximization of production and quality. The other factors that aided management to be the basic tool for handling any activity are the globalization and information technology with all its ramifications and exciting consequences for success in competition for organizations, groups and individuals as well and education being no exception.

10. Systems Theory

The definition for the system may be taken as that as a set of interrelated parts that function in tandem as a whole to achieve a common purpose'. Systems concept has broad meaning and systems will have boundaries and they also interact with environment and the organization. In this aspect management treats organizations as a set of inter dependent systems depending upon the area of operation and the nature of the production activity be it a product or service which are inter connected in achieving the goal or objective.

Systems theory is synonymous with the concepts of entropy – the tendency for a system to run down and die, and synergy – the concept that the whole is greater than that of its parts and subsystems parts of a system that depend on one another for their functioning.

Another important feature of system is that there are inputs and through a process, output is obtained. The presence of internal and external environment and the important process is the feedback between input and output are the important features in systems theory.

11. Mathematical or Management Science Approach

In the quest for precision and exactness in management and its approach the methods of mathematics, inductive, deductive and symbolic logical principles are being utilized wherever they are applicable and able to find solutions for problems in management. Even though most of the problems can be modelled on mathematical basis there is a serious constraint and limitation in applying this approach to each and every management problems that is not feasible always. This approach handles the problems in cause and effect and contingency aspects.

12. Roles Approach of Mintzberg (1939)

This approach was proposed by Mintzberg after observation of five chief executives of different organisations. He opined that managers have three basic and fundamental roles i.e., 1) interpersonal 2) informational and 3) decision roles. The limitations for this system are the sample taken for study was very small based on which the conclusions are drawn. Some activities are not of management as has been claimed. Some important activities are left out or omitted like appraising managers.

13. 7-S framework approach

Initially the 7-s framework is proposed by Waterman and Peters and was taken forward by Mackinsey by using seven concepts all starting with the letter 'S' and making a management model to explain its role. They are 1. Strategy, 2. Structure, 3.Systems 4 .Style 5.Staff 6.Shared Values and 7.Skills. The system proposed by Mckinsy does have some resemblances with the theory proposed by Koonz et.al. since the year 1955, and also the terms were not explained in depth nor with precision.

14. Operational Approach

This approach draws heavily from other fields of knowledge, the concepts principles and techniques and also the content or knowledge of management approaches, it has clearly drawn a differentiation between managerial and non managerial knowledge. It also has facilitated in developing the classification of the management system based on managerial functions viz. planning, organizing, staffing, leading and controlling. However the major weakness of this approach happens to be that of not identifying or representing the vital function of 'coordination' and its purpose in management.

As the problem taken up for investigation is interdisciplinary it became a basic prerequisite to review the concepts in the branches of knowledge of philosophy with a stress on ethics and management and education.

In the next Chapter a review is attempted on the studies conducted on education and the allied aspects of the present topic both Indian and foreign contributors.

"The secret of education lies in respecting the pupil."

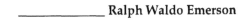 Ralph Waldo Emerson

Chapter – 2

Abstract

This chapter deals with the relevant literature with reference to the various studies that are carried out in the field of education both by Indian and foreign researchers with a special reference to higher education in India. An attempt is made to cover the most relevant aspects of higher education which can influence the process of higher education in different dimensions and directions. Issues like the role of society, policy and programmes their interaction, synergy and dynamics in higher education in India are covered.

REVIEW OF LITERATURE

"Morality is the best of all devices for leading mankind by the nose."

_____ Friedrich Nietzsche

For better elucidation in this regard, research studies conducted in the past as well as in the contemporary period have been presented. Further, in order to have comprehensive understanding on the trends of education and related issues the research studies were categorized on the basis of social and economic, personal traits etc., influence on education achievement in the educational achievement, policy and impact studies etc. were presented. Research Studies conducted in other countries and also studies of foreign authors were also been presented.

The chapter also provides the contemporary situation of the problem chosen for research investigation. As cited by many a author review of related literature is an important pre-requisite to the actual planning and then execution of any research work. John Best (1982) summarises the importance of review of relevant literature in the following manner:

"A summary of the writing of recognized authorities and the previous research provides evidence that the researcher is familiar with what is already known and what is still unknown and untested. Since effective researches are based on past knowledge, this step helps to eliminate the duplication of what has been done and provides useful hypotheses and helpful suggestions for citing studies that shows substantial agreements and those that seem to present conflicting conclusions help to sharpen and define understanding of existing knowledge in problem area, provides a background for a research project."

True to the observations of John Best, the review of literature immensely help in leading to comprehensively understand the core issues of research problem to be investigated. In fact, the review of relevant literature leads to understand the past work undertaken in this regard and remove prejudices if any. The review of literature also leads to sharpen the analytical insight on the research study outcome so as to reach logical conclusions and remove bias if any. This in background, the following is the brief review of the relevant research studies undertaken which were presented under different categories.

I.   Studies on Education in General

In this section an attempt was made to present the outcome of research studies conducted earlier with reference to importance of education, its impact on various issues of the society and other related issues. These studies were presented in two distinct sections of studies conducted in foreign countries as well as the studies conducted in Indian context.

a.   Foreign Studies

Clark (1927) found that students whose parents had college education ranked higher in scholarship. Shuttleworth (1927) reported the low-achieving group of students had strict religious home training. Bear (1928) found parental occupation related to academic success. He reported that sons of farmers and businessmen ranked low in scholarship in comparison to those of artisans, salesmen and so on.

Austin (1974) found very high relationship between the tendency to drop out of college and parents' education and father's occupation. Sinha (1970), and Wig and Nagpal (1970) found low achievers represented more in occupational category agricultural or business. Griffits (1926) found a close relationship between school grades and family size. Children from small families were found superior in school grades.

Havighurst (1964) contrasted achievement test performance of middle-class and lower-class children in twenty one (21) Chicago school districts. He found that sixth (6) grade students in the seven (7) districts with the highest average of socio-economic status ranged from a grade level to one year above grade level on reading and mathematics tests; in the seven lowest socio-economic status districts, the plus one (+1) scores clustered around one year below grade level.

b. Indian Studies

Mishra, Dash and Padhi (1960) reported a correlation of 0.59 between home environment and school achievement whereas correlation of 0.31 between intelligence test scores and school achievement. Menon (1973) found over-achievement and under-achievement are highly influenced by socio-economic status. Anand (1973) established relationship between socio-economic status and academic achievement even when the influence of intelligence of non-verbal and verbal type was partialled out. He also found that the impact of socio-economic environment was found to influence mental abilities and academic achievement.

Abraham (1974) found achievement level in English is associated with socio-economic status and Basavayya (1974) observed overall language achievement is influenced by the parental occupation and education. A study on difficulties in learning English by Dewal (1974) revealed that effective teaching and learning are hampered by poor socio-economic background. Bhaduri (1971) observed that the over-achievers showed higher scores on study habits, attitude to school, and religious-cultural background; the under-achievers on the contrary, tended to have a higher socio-economic status, a more congenial home condition and more of leisure time activities.

Lalithamma found that the achievement in mathematics was positively related to intelligence, study habits, interest in mathematics and socio-economic status. Correlation between, socio-economic status and academic achievement as computed by Prakash Chandra (1975) was reported as positive and is supported by Homchandhuri (1980), Khanna (1980), Shukla (1984), Shukla (1984) Mehrotra (1986), Misra (1986), Singh (1986) and Rathaiah and Rao (1997) also. Satyanandam (1969) highlighted two sub-aspects of socio-economic status, viz., educational level of parents and economic status of parents. According to him the children of graduate parents performed far better than the children of matriculate parents.

Children of upper and lower, upper and middle economic strata only differed significantly on the variable of achievement.

Chatterji, Mukherjee and Banerjee (1971) also found that parent's education level was directly related to the achievement of their children. Khanna (1980) observed that the academic achievement of the children of educated parents, illiterate parents, and educated mothers was significantly correlated with the socio economic status of the family. Menon (1972) also noticed that higher occupational and educational level of father, educational level of mother, family income and parental attention were related to high achievement.

Ojha (1979) concluded that the higher the socio-economic status, the better would be the academic achievement at high school level. Parental education, occupation, and income were also related with the educational achievement of both rural and urban boys of ninth (9) class. Choudhari (1975) expressed his opinion based on research that bright children normally come from families where parents having a higher level of education, were mostly engaged in professions requiring general knowledge, and had more income than the parents of dull students.

In Goswami's (1978) study, the scholastic achievement correlated highly with socio-economic status. Goswami (1982) found a significant relationship between socio-economic status and reading interests and also between reading interests and academic achievement.

Jain (1981) states that, the socio-economic level of the parents had a great impact on the pupils' achievement in Gujarati, in the subjects' social studies, science and mathematics. The pupils belonging to the upper socio-economic status achieved better than the pupils whose parents belonged to the middle and lower socio-economic levels, while the pupils from the middle socio-economic levels scored better than those with lower socio-economic status of the parents, in all the subjects. Academic achievement had a high positive correlation with socio-economic status.

Family background factors of college students, according to Siddiqui (1979) had positive relationship with the academic achievement of the students when the intelligence factor was held constant. Griffits (1926) observed that within the family the older and the younger children tended to perform about equally well scholastically. Gupta (1982) found that birth order and the father's profession influenced the reading ability (in Hindi) of children studying in classes. Chatterji, et. al (1971) concluded that the family size and the number of siblings were inversely related, especially in low intellectual level. Dave and Dave (1971) observed that the size of the family was not related to the academic achievement. Dave and Dave (1971) noticed that a higher percentage of rank students belonged to homes having parental income, occupation and education, whereas a higher percentage of failed students belonged to homes having lower parental income, occupation and education.

In the study of Dhami (1974) the relationship between socio-economic status and academic achievement, though statistically significant, was not very high. Socio-economic status was moderately correlated with achievement in the study of Srivastava (1981). Sinha (1970) also observed only small differences on their parents' education and father's education. The socio-economic status of the pupil's parents was not significantly related to scholastic performance at Class VIII and Class IX but at Class X the pupils hailing from homes with higher socio-economic status performed better.

Nomzek (1940) reported that education of parents and their profession have no influence over the academic success of their children. But for the high ability group, children of servicemen excelled the children of businessmen, and the trend was reversed for the average and low intellectual groups. Salunke (1979) found no relationship between socio-economic status and achievement. Bhat and Indiresan (1981) failed to draw definite conclusions regarding the differential performance of students belonging to different socio-economic backgrounds as the sample consisted mainly of students belonging to the backward class and low income group.

Chatterji, Mukherjee and Banerjee (1971) concluded that the economic conditions of the family seemed to have no effect upon the scholastic achievement in all the intellectual ability groups. They also found that father's occupation was not consistently related to children s achievement. Desai (1979) observed no relationship between socio-economic status and achievement.

Shukla (1984), Mehrotra (1986), Misra (1986) and Singh (1986) showed a positive relationship between Socio-economic Status and academic achievement of the students. In the study conducted at the CIII. Srivastava and Ramaswamy (1986) found that the effect of Socio Economic Status on achievement in mathematics and social studies was significant. Dwivedi (1983) found that Socio-economic Status significantly affected achievement in Biology of higher secondary pre-medical students when taught through a linear programme. In the study of Sarah (1983) it was found that the coefficient of correlation between achievement and Socio Economic Status was positive and significant when the effect of pupils' attitude towards social and science education were partialled out.

In the study by Das (1975) which was conducted in West Bengal, the socio-economic status was one of the primary factors responsible for low achievement in general science. Studying the relationship between certain psycho-sociological factors and achievement of student-teachers in teacher training institutes of Andhra Pradesh, Goplacharyulu (1984) showed that socio-economic status and caste influenced the total achievement as well as achievement in theory and practical, taken separately, of the student-teachers.

Pandey (1981) and Puri (1984) studied the influence of environment as a factor to promote academic achievement among students. The former concluded that an urban atmosphere was more conducive to better achievement than a rural environment. The latter brought out that the effect of environmental facility on both general academic achievement and achievement in English language was significant.

Studying the effects of home environment on the cognitive styles of students, Paul (1986) concluded that the factors of home environment, like recognition of the child's achievement, parental aspiration, forbearance for the child's wishes, parental affection, encouragement for initiative and freedom, etc., had positive and significant correlation with each of the four modes of cognitive styles studied. Grover (1979) indicated some influence of aspirations of father and mother over children's academic achievement. Gaur (1982) showed that birth order did not affect the speed of reading, comprehension and vocabulary of students. Trivedi (1987) found that parental attitude was significantly related to academic achievement. Lall (1984) showed that protective attitudes of parents positively related to the academic success of boys. Jagannadhan's (1985) study indicates significant effect of home environment on academic achievement.

The environment provided at the learning place of students as a variable has been studied by Deshpande (1984), Doctor (1984) and Upadhyaya (1982). No specific trend of organizational climate was found to differentiate between the high and low achieving schools, according to Deshpande (1984). The study by Doctor (1984) indicated a relationship between classroom climate and academic achievement. Upadhyaya (1982) conducted the study on the tribal population of Bastar District in Madhya Pradesh. It was found that each of the three aspects of classroom environment-interpersonal relationships, goal orientation, and system maintenance and change-was significantly related to academic achievement.

Studies by Girija (1980), Mishra (1983), Malik (1984), Kamila (1985), Pandey Kalpalata (1985) and Verma (1985) have concentrated on samples of students who may be considered as slightly disadvantaged when compared to others. Kamila (1985) brought out a comparative picture between the achievement of students belonging to Harijan and Tribal Welfare Department high schools and those belonging to Education Department high schools in Orissa. The picture was in favour of the latter.

II. Education: Role of Society, Policy Issues and Programmes

In this broader section research studies conducted on various important aspects related to community participation and its role; Studies on various Programmes and Initiatives from government sector as well as society; Impact of education on various issues related to society and its emphatic role there of etc. have been presented. Ramachandran, Vimala (2003) in their research titled "Backward and forward linkages that strengthen primary education" studied in-depth the aspects of backward and forward linkages in strengthening primary education. It is widely acknowledged that a significant portion of children from underprivileged backgrounds either drop out before they reach Class V, and if they continue to attend school, learn very little. . There is a wide gap in learning achievements between government schools and private schools.

The universalisation of elementary education would not be possible unless three important areas are addressed viz., pre school education, remedial education and bridge programmes for children who dropout and post primary education. The active participation of children in primary education depends upon the factors like physical access, dysfunction of schools, motivation and commitment of teachers. Under this backdrop, case studies of two organizations M.V. Foundation in Andhra Pradesh and The Concerned for Working Children in Karnataka have shown the way and the larger education community has a lot to learn from them. There are other set of case studies mentioned in the paper are: Pratham of Mumbai,

Nali Kali in Mysore, Digantar in Jaipur of Rajasthan, Agra Gami experiment in Orissa, Muktangan experiment in Rajasthan.

One common short coming is non availability of quantitative data and long term sustainability. Economic prosperity has improved educational aspects especially for girls in middle income families. But the situation of poor girls of below poverty households is a cause for concern so is the case of children from SC and ST communities. Lessons drawn from the case studies should pave the way to improve pedagogic renewal, teacher motivation and training, community mobilisation, activating mother-teacher association, altering school environment, reworking text books to make them gender sensitive and with incentive and support schemes. The creation of forward and backward linkages is essential to create an environment where every child not only goes to school but benefits from it.

Varshney, Hemant Kumar (2002) in the study based on secondary data titled "Inter-state gender disparity in literacy rates: A look at census data (1991 & 2001)" analyses the trends in gaps in male-female literacy rates, both at national and state levels. The data drawn from 1991 and 2001 Census of India is used for this purpose.

It is observed that for the country as a whole the literacy rate in 2001 for the total population, males and females is 65.38 per cent, 75.85 per cent and 54.16 per cent respectively i.e., nearly one fourth of males and half of the females in the country are still illiterate. However, the literacy rate has recorded a significant jump of 13.17 percentage points from 52.21 in 1991 to 65.38 in 2001.

The increase in the literacy rates of males and females for the same period is 11.72 and 14.87 percentage points respectively. A comparison between male and female literacy shows that annual compound growth rate (ACGR) in literacy among females is significantly higher than males. During 1991-2001, the annual growth rate in literacy rates for females is 3.26 per cent as against 1.69 per cent for males. The gap in male-female literacy rate is 18.30 percentage points in 1951 which has increased to 26.62 in 1981. It has shown a declining trend in 1991 and 2001 from 24.84 and 21.19 percentage points.

At the state level also, though there is improvement in literacy rates for both the sexes over the period of study, male literacy is found to be higher than female literacy. Kerala with a literacy rate of 90.92 per cent holds the first rank whereas Bihar with 47.53 per cent ranks last in the country. It is concluded that there is an improvement in male-female literacy rates and it contributed significantly in narrowing down the gender disparity.

Garg (2002) made an attempt to understand the role of state as well as civil society in promoting education during the decade of 1991-2001. The study examines the progress of literacy in India especially during the decade 1991-2001 and analyses the role of the Government and civil society in the improvement in literacy rates. It is observed that the First and Second Five Year Plans of 1951-56 and 1956-61 provided for social education, including literacy for adult population of the country. The literacy rate increased from 18.33 in 1951 to 52.21 per cent in 1991 recording an average decadal growth rate of only 8.5 percentage points. The literacy rate progressed by

18.17 percentage points during the decade 1991-2001 which was about 1.6 times more than the decadal average of the last four decades. Eradication of illiteracy was one of the major concerns of the state policy.

National Literacy Mission (NLM) was set up in May, 1988 to adopt "mission" approach in combating illiteracy. Total Literacy Campaign (TLC) for different districts was envisaged. As a result of sustained efforts at all levels by 2001 the literacy campaign reached five hundred fifty nine (559) out of five hundred eighty eight (588) districts of the country. Eighty four (84) million persons have been made literate as a result of the adult literacy programmes launched by NLM. Rajasthan, which was having a literacy rate of only 35.55 per cent in 1991, improved the same to 61.03 per cent in 2001. Chattisgarh also gained in the literacy rate by 22.27 percentage points from 42.91 in 1991 to 65.18 in 2001. Literacy rates of Madhya Pradesh also improved by nineteen (19) percentage points. Except Bihar, the other low literacy states of the Hindi belt had improved their literacy rats. Civil society played an important role in raising the issues of disparity in providing educational facilities to women and weaker sections. Several NGOs and women's organisations also contributed in improving the literacy rates among women.

Lhungdim (2000) examined the aspirations of adolescents in Manipur and assessed their implications for educational planning in the state. The study titled "Aspirations of Adolescents and their Implications for Educational Planning: A study in Imphal and Churachandpur Town, Manipur" examines the kinds of aspirations the adolescents in Manipur have and relates their importance as a tool for policy making in the field of education. Data were collected from 1655 adolescents from fifteen (15) schools in Imphal and nine (9) schools from Churachandpur town of Manipur State during 1996-97 through Survey.

Study reveals the interplay of many factors, and the kind of hardships adolescents have to undergo to achieve what they aspire for. Most adolescents in Manipur, inspite of being talented, may not be as fortunate as their counterparts in other states when it comes to the question of pursuing higher studies. This calls for an urgent need to realize and restructure the linkages between educational system and other socio-economic, political and cultural factors. Only less than eight (8) per cent of the sample adolescents aspired for the teaching profession despite the fact that a very good opportunity is available for this job within State. Students in a State like Manipur need facilities such as counseling etc. much more than in any other state not only in terms of their future prospects but also to cope with the unhealthy academic and socio-political environment prevalent in the State. Majority of the students preferred to study in schools than government sponsored. Aspirations of the adolescents reflect their interest in a particular subject or profession like doctors, engineers, army officer, etc. Most of the parents of the adolescents especially their fathers have a stable job with educational level ranging between higher secondary and graduation. At least one half of them preferred to study anywhere outside the state. There is a need to promote career counselling in each school, at least a year before students complete their schooling, because substantial proportion of adolescents have no aspirations for the future at all. It is concluded that there is a tremendous responsibility on the part of the parents and the state to properly channalize the aspirations of adolescents lest they should go astray and mar their future.

III. Studies on Higher Education

In this section an attempt was made to understand the outcome of research studies conducted in the realm of Higher education in India.

VK Patil (2000) in his publication titled "Studies in Higher Education" made a comprehensive attempt the issues pertaining to Higher Education. He opined that "Higher Education in India has come in for much attention in recent years and the reasons are not far to seek. Several issues like equity vs. excellence, privatisation of universities, the growth of self-financing colleges, the concept of university education as a "non-merit" good, and the emergence of open universities are increasingly getting into focus. This book is yet another attempt, quite welcome at that, to gather the views of educationists on many topics of interest in this field." The book edited by him has forty (40) chapters arranged under four parts deal with a variety of issues, some of them sensitive that pertain to several recent developments. Here is a sampling of the fare that is provided: Prof. K.B. Powar traces the growth of higher education in the country since Independence and pleads for a convergence of the formal and Open University systems. While admitting that the UGC's scheme of vocationalising first degree education is a "bold initiative which should give a new direction to the relevance of higher education", Prof. M.V. Pylee urges for greater attention to the linkages between work experience and theoretical knowledge, and employability and relevance.

The crucial importance of moral education cannot be denied and Dr. V.T. Patil and Dr. Narayana treat this topic in an objective fashion, while discussing "Higher education and social transformation." External Quality Assurance (EQA) has come to be accepted in many countries as a measure to maintain and sustain quality in universities and colleges and Dr. A. Gnanam outlines the way this mechanism has taken shape in India. Dr. V.S.P. Rao enumerates the varieties of factors, which go to render university administration a complex task. He regrets that many universities have not recognised the value of long-range planning and deplores the overemphasis on paper work, rules and regulations.

Whether higher education should continue to be subsidised by the Central and State Governments has been a ticklish subject all along and Dr. Marjorie Fernandes discusses the salient features of the publication on "Government subsidies in India." The author builds up a rationale for subsidies to be given to higher education. This matter claims attention by three more authors in subsequent chapters. Urging that our educational programmes, curriculum and syllabi must be such as to meet national needs and global challenges, Dr. V.C. Kulandaiswamy (2006) makes out a case for remodelling some existing organisations like the University Grants Commission (UGC).

Ashokan and Virk (2004) have highlighted the plight of higher education in India. Mushrooming growth of self-aided colleges, deemed universities, colleges with autonomous status and the rest have not come to the rescue of the student community to achieve what it rightly desires and deserves. Perhaps, the order of the day is to trade on subjects with different catchy names, enticing the plethora of students. Distance education is a massive factor in the dilution of higher education. Clearing UGC/CSIR/NET is tough because the syllabus framed for a subject by different universities is not the same. UGC alone can serve as a placebo and warrant for a uniform syllabus all over India for any specified subject for the benefit of students. Teachers should not be promoted based on the completion of orientation and refresher courses that they have attended. An assortment of factors like performance appraisal, projects undertaken, research activities, publication of papers, symposia or workshops attended could be used as tools for promotion. By doing so, UGC can also conserve its resources only to channalize them for other useful activities. UGC can organize orientation courses on a routine basis for all candidates aspiring to enter the portals of higher education and emphasize it as a pre-requisite for appointment. It is high time UGC revamps the entire system of higher education to cater to the needs of the student community.

Powar, KB (2000) through his work on "Indian Higher Education - A Conglomerate of Concepts, Facts and Practices" made an attempt to elucidate on intricacies of higher education in India. The title of the book under review is in a way tantalising but the author clears the mystery in his preface. A conglomerate is a rock unit made of well-rounded pieces of rocks of different compositions, cemented together by a matrix of any material. As a geologist and educational administrator of long standing, the professor has used the term "conglomerate" — the commonly shared concern for higher education being the cementing matrix of the different chapters. And he has admirably succeeded in the task.

Indeed, the contours of the fascinating topic are well delineated in five sections — Concepts and development, The Indian context, Financing higher education, In search of quality, and Distance Education. The twenty one (21) chapters (the rocks in geological parlance), which are encompassed in these sections, give an authentic and lucid picture of the scene.

Prof. Powar has not shied away in writing bluntly about certain recent events. For instance, he says "A disturbing development of the 1990s has been the influx of foreign universities that prefer to enter into partnership with professional organisations and little-known institutes that do not form part of the Indian higher education system... The activities of such universities need to be discouraged. The Association of Indian Universities, on its part, has not recognised the degrees offered by foreign universities through programmes in India." This is a bold and authoritative stand since the author is also the secretary-general of the Association of Indian Universities.

That Distance Education (DE) has emerged, as a viable and effective option to a large section of youth having no access to the conventional universities, is well brought out by the author. Only about six to seven per cent of those in the age group of 17-23 (abysmally low when compared to developed countries) are in the conventional stream, but the silver lining here is the way in which DE (with ten (10) open universities and over sixty (60) regular universities offering correspondence courses) is making "access and equity" a realisable dream.

Similarly, writing about the universities of the 21st century, the author brings to bear on the subject a refreshing insight. "Imparted under the all-pervading influence of the communication technology revolution, it will be student-centred, committed to the concept of life-long learning, responsive to the needs of society, increasingly privately financed, and influenced by market forces. It will be international in character placing emphasis on quality, with partnerships and networks being important."

Although the Indian higher education system has several deficiencies, there are positive aspects as well, and the author has listed them – starting the special assistance programme under which selected departments in universities get support for research; establishment of curriculum development cells for developing model curricula at both undergraduate and post-graduate levels; granting of autonomy to selected colleges so that innovations in the academic field become possible; establishment of academic staff colleges for imparting orientation courses (for new teachers) and refresher courses (for serving teachers).

Kaul, Rekha (1993) in her work on "Caste, Class and Education: Politics of the Capitation Fee Phenomenon in Karnataka" made an in-depth study on policy issues pertaining to higher education in Karnataka. The book is divided into eight chapters. The first two chapters deal with the historical aspects of the establishment of private colleges and the emergence of the phenomenon of capitation fee

The most interesting chapters are three and four which discuss the growth of the capitation fee phenomenon and the government policy towards such colleges. To gain political influence, governments of Karnataka irrespective of political ideologies encouraged the establishment of private colleges mainly motivated by profit, by organizations and individuals.

In conclusion, the author suggested some specific measures to improve the quality of education in the private colleges. Some of them like sponsorship of students by industries and opportunity to work there during their education are useful. However, the hope of the author is, "but once the demand for quality education is widespread there would be checks on such private colleges."

Kulandai Swamy, C (2006) in his work on "Reconstruction of Higher Education in India" made an exhaustive understanding on the higher education in India and the way forward. He opines that after the introduction of neo-liberal reforms, various sectors of the economy have grown at a rapid pace, generating great demand for trained manpower. This situation presents both a challenge and an opportunity, but as educationalist V.C. Kulandai Swamy narrates in this crisp overview of higher education, little has been done to build a solid framework for human resource development in India.

Kulandai Swamy, (2006) who has served as Vice-Chancellor of general, technical and open universities, draws from his considerable experience as an academician and educational administrator to document the ills that shackle what should otherwise be a vibrant education system; he then proposes a way forward. Among the major stumbling blocks to development, he thinks, is the university affiliating system for colleges. He argues that it should be dismantled without delay in favour of universities and colleges that vary in size, student strength and academic offerings. They must also adopt the semester system and credit-based course pattern.

The book is a broad sweep with a macroscopic perspective, providing statistical analyses on India's attainments in the educational sphere. It encompasses not just the university system, but discusses the prospects for orderly growth of higher education with the active involvement of the private sector. There is an interesting, although brief account of how the Centre did not pursue legislation that would enable formation of good private universities. Poor preparedness to face international competition in provision of education as envisaged by the General Agreement on Trade in Services (GATS) is also similarly exposed.

The mediocre quality of higher education is analysed candidly, with an expose of the weak foundations of the system. It is revealed, for instance, that only a third of the colleges in the country fulfilled the minimum requirements specified by the UGC for recognition, during 2003-04. In the case of postgraduate courses and research degrees in most non-technical subjects, there is a climate of laxity and insufficient academic rigour.

Part of that problem, according to the author, can be traced to the low scale at which institutions operate. Many international universities of repute, by contrast, have large campuses and several thousand students. They are richly endowed with talented faculty and intellectual pursuits thrive in this milieu. A reconstruction of the framework of education would encourage such universities and deemed-to-be universities to come up in India.

The author traces the lack of initiative relating to expenditure on education as a percentage of national income. He emphasises the need for investment of at least 5.5 per cent in the Eleventh Plan. The reality of higher education becoming an economic good that must be paid for is also acknowledged.

There is of course a case for a longer discussion on equity in access to all in education when a significant section of the population suffers extreme deprivation. Can all sections purchase the economic good that education has become with their meagre resources? Should they be compelled to pay high fees through costly loans, or can they hope to benefit from recognition of entitlement and from endowment models based on community participation? This book presents material to take that debate further.

Mohan Ram (2004) in his work on "Universalisation of Higher Education: Some Policy Implications" made an examination of pattern of higher education prevailing in India and analysed the implications there of and later on presented an exhaustive way forward. In his opinion - "Education aims to transmit aspects of human experience to the as yet uninitiated, and so involves an introduction to modes of behaviour - which certainly include modes of thinking and feeling-relevant to our common life. Although many impressive achievements have been made in higher education since independence but a closer analysis reveals that situation is still gloomy and very far from satisfactory. More than half of our population is still illiterate. Adult education is far from its targets. The goal of a universal education has not been obtained. Education of girls has been lagging behind that of boys. Wastage of resources due to dropout and stagnation still continues especially in females, SC/ST and socially disadvantaged sections for our society, especially rural areas. The absence of adequate information base and proper infrastructure has a series of adverse effects on overall educational planning.

There are some definite linkages between educational development and disparities. Tackling as it does varies concerns which are of growing importance in most developing countries, the collection in this book is of thought provoking critical reviews/papers/articles from India and abroad which would appeal to a wide range of readers." (jacket)

Niyati Bhat (2004) in his work on "Higher Education Administration and Management" studied elaborately on the issues of: Administrative organisation in higher education; Theories of administration; Bureaucracy in higher education; Allocating academic resources in higher education; Systems planning in higher education administration; Learning resources in higher education; Personnelling in higher education; Faculty and staff evaluation process in higher education etc.

Niyati Bhat opines that "The administration of institutions of higher education is a very complex, challenging, and, in many instances, frustrating undertaking. The administrator must deal with many groups, among them students, faculty, other administrators, federal, state and local governing agencies, accreditation agencies, business and professional organisations, service clubs and alumni. In order to effectively manage this task, today's administrator must be thoroughly familiar with the various ground rules, regulations, and laws that pertain to higher education.

The book is covering administration, theories, women administration, academic resources, bureaucracy and learning resource in higher education. The book is prepared for teachers, educators, researchers and for those which are concerned with the administration of higher education. Madan, VD (2002) through his work on "Higher Education Beyond 2000: An Omni Tech Approach" touched upon various facets of higher education in India. He emphasized elaborately on pedagogical issues, training etc.

Madan emphasizes that "Technology-explosion, knowledge-globalization and education-massification are the three most dominant factors forcing the convergence of the two mega-systems of higher education: conventional education system and distance education system, leading to the emergence of an inevitable and innovative Omni-Tech Education System for the new millennium. This sequel is a venturous exposition of a technology-based distance learning approach of higher education beyond 2000."

Venkataiah. S (2001) brought an edited volume on higher education in India with focus on various facets of the same. However, he elaborated more on use of technology in higher education. He argues that "few professionals in higher education today would dispute the importance of the effective use of

Review of Literature 74

technology in teaching and learning for the overall success of an institution."

Venkataiah further emphasizes that "many vital problems and related issues are discussed here in this book. The major among these are—practical problems in higher education; management information systems; artificial intelligence; learning contracts; higher education and interactive video-disc; workplace learning; teaching and learning; interactive multimedia; higher education of women; education and economy; information technology; integration technology into teaching and learning; and computer literacy course." Various articles presented in the book is in fact elucidate the readers on latest knowledge in the field.

Oza, N.B and Joshi, K.M (2001) brought out a work on Higher education in India with the title "Higher Education: Issues and Options." The book is a collection of works on issues like industrial support for higher education: major impediments and policy prescriptions, Self-financed educational programme in university system, Relevance and necessity of self-financing in higher education, Total Quality Management (TQM) in higher education, Academic staff development in higher education, Higher education for women: experiences of India and other countries, Examination: a gravitational force of education and Teacher's motivation: role, problems and remedies, Examination reforms in higher education.

Oza (2001) emphasizes that "Higher education is of great significance for the balanced development of an economy and shall definitely have its impact, especially for a developing country like India. Today, the system as a whole of higher education in India has grown a lot. The number of institutions has increased manifold. Enrolment in universities has also increased over the years, although still far below that in developed countries, and even less than in the South Asian countries.

The responsibility of providing able and well-trained graduates is that of higher educational institutions of India. Higher education is inextricably linked with industry and the competitiveness of industry is determined by faster exploitation of new knowledge and technology. Therefore, the author believes that it falls upon the government and major industries to sponsor higher education in the country. With a perpetual resource crunch in the higher educational institutions, industrial support is naturally sought after. By this, the industries are sure of obtaining graduates trained in specific, state-of-the-art technology, to be absorbed in their firms.

Eminent authors in the field have contributed articles in the book. Aspects like self-financing and TQM in higher education have been dealt with authoritatively. The book will certainly be of immense value to all institutions of higher learning on the one hand, and to HRD and personnel executives of major industries on the other hand, providing them with the justifications for sponsoring higher education."

Bharat B. Dhar (2008) brought out his work on "Higher Education System." elaborated on issues like: Higher education scenario in the twenty-first century: accepting challenges to change; Quality and accreditation: complementary to each other; Teachers and their professional development; Supervision and research in the universities: some issues and problems; Suggestions on scientific and technological research in Indian Universities; The concept of research methodology; Autonomy within a university system; Management concept of higher education system and the need for modern tools like computerization and information technology; The importance of ICT in teaching: is it growing; Environment education in India: status and strategies and Internationalisation: to what extent?

Bharat D. Dhar opines that "Higher Education System no doubt is complex, but in recent decades it has gone through a sea change through its curriculum development, interdisciplinary approach and internationalization. Any university of the country is judged by the level and the extent of research work it carries besides the regular teaching. The quality of teaching imparted to the students is enriched by the research its faculty members carry through. The modern concept of knowledge society in higher education system has gained significant importance in its all dimensions. In this book attempt has been made to consolidate together a selected set of articles based on issues of relevance to higher education. This document though has only a few selected articles by the author, will be an effective tool in the hands of higher education community.

Focusing on the issues of higher education, especially issues like education and political economy, quality of learning etc., V.K Rao (2003) brought out his work on higher education titled "Higher Education." V.K Rao elaborated on issues like: Interactive higher education; Teaching and learning; Problems in higher education;

Higher education of women; Education and political economy; Modern education; Educational policies in India; Basic education; Technical skills of teaching; Quality of learning; Quality teaching; and Challenges in higher education.

V.K Rao opines that "Higher education is too serious a matter to be left wholly to politicians and bureaucrats. Our leaders, both at national and state levels, have generally been aware of the need to keep higher education away from political interference. Right from the time when India attained independence expert committees have been constituted and periodic reviews made about the needs and problems in the field of higher education.

P.K Nayak (2002) made an elaborate study on higher education status in Arunachal Pradesh. Since Arunachal Pradesh, by virtue of its remoteness and geographical constraints has been facing quite a number of problems in promoting higher education. In this context, his work is quite significant. P.K. Nayak reflects the status of higher education in the following manner:

"Education, almost universally, is seen as a powerful instrument making positive and effective contributions for a better world. It is such a large undertaking and it has so radical an influence on man's destiny that it can be proved damaging if it's very substance, its relationship to man and his development, its interaction with the environment, as both product and factor of society are not deeply scrutinised and extensively reconsidered. In a country like India where variety marks physical, social, economic and cultural aspects, any worthwhile attempt at understanding the way education operates at the national level must endeavour to deal

Review of Literature 78

with variations peculiar to local contexts. The study of a micro-level unit and the like put together relieves such apprehensions.

The present work reports the research carried out on one such phenomenon, viz. Need for higher education in Arunachal Pradesh, basically stressing upon the dropouts. This volume focuses on the following various aspects of education—primary, secondary and higher. Certain remedies for the eradication of various maladies are also suggested. The volume grown upon many tables and analyses is expected to be popular among the readers and researchers on education of Arunachal Pradesh."

Sensitizing the issues related to higher education and the challenges to be faced in new millennium, R.N. Dhir (2002) brought out an elaborate study through his work titled "Higher Education in the New Millennium." Given the focus of the study, R.N. Dhir concentrated on issues like Educational opportunity in higher education; Higher education: teaching and learning; Objectives of higher education in a changing world, Financing of higher education; Impact of management information system; Interactive education; Open University system; Pertinence and quality; Higher education in India; and Future of higher education.

While emphasizing on the application of technology in higher education in the new millennium, R.N. Dhir expresses that "The revolution of Information Technology is bound to have a profound effect on the role of higher education in India. This book presents the most recent ideas and findings about the higher education. The volume emphasizes new perspectives on development and on technical analysis. This authoritative and comprehensive text is a landmark in higher education the focus is sharply on research educational opportunities, teaching and learning, interactive education, Open University system etc. In most of the chapter the approaches and methods are reviewed by studies from higher education."

The above is the exhaustive presentation on research works attempted in the realm of higher education and associated fields. As may be observed, though elaborate studies were conducted and published through various media, the research has been largely confined to issues related to Higher education and its spread in the changing circumstances. The challenges of higher education were largely associated with the issues of policies, programmes, geographical issues, and perception of students, application of technology and other related aspects. As such, the environment and ethical issues related to higher education management were not much attempted.

The above in view, the present study is taken up.

In the next chapter an attempt was made to cover the higher education in India and the various arrangements institutions, administration, governance and setup and structural aspects.

"The highest result of education is tolerance."

- Helen Keller

Chapter – 3

Abstract

This chapter deals with the system of higher education in India, the structure, process, different land marks that govern the system from time to time. It is also elucidated with reference to objectives, goals, demands, problems, planning and execution of higher education. The role of different higher educational institutions in different fields and their contribution and their journey towards the fulfillment of the mission of imparting education to the demands of the society and their contributions were discussed.

HIGHER EDUCATION IN INDIA

"We should not value education as a means to prosperity, but prosperity as a means to education. Only then will our priorities be right. For education, unlike prosperity is an end in itself. .. Power and influence come through the acquisition of useless knowledge. . . irrelevant subjects bring understanding of the human condition, by forcing the student to stand back from it."

_____ Roger Scruton.

Higher Education in India has come in for much attention in recent years and the reasons are not far to seek. Several issues like equity vs. excellence, privatisation of universities, the growth of self-financing colleges, the concept of university education as a "non-merit" good, and the emergence of open universities are increasingly getting into focus. Whatever the case may be, an attempt was made in this chapter to understand the Higher Education in India.

Higher Education in India is one of the most developed in the entire world. There has in fact been considerable improvement in the higher education scenario of India in both quantitative and qualitative terms. In technical education, the Indian Institute of Technology's (IIT), and in management, the Indian Institute of Management's (IIM) have already marked their names among the top higher educational institutes of the world. Moreover the Jawaharlal Nehru University and Delhi University are also regarded as good higher educational institutes for pursuing postgraduate courses and for research in science, humanities and social sciences.

As a result, students from various parts of the world are coming today for higher education in India.

i. Higher Education system in India:

Higher education in India starts after the higher Secondary or 12th standard. While it takes 3 years for completing a B.A., B. Sc or B.Com pass or honors degree from a college in India, pursuing an engineering course would take four years and five years (with six months of additional compulsory internship) for completing a bachelor of medicine or bachelor of law degree. Postgraduate courses generally are of two years duration. But there are some courses like Master of Computer Application (MCA) that are of three years duration. For those who cannot afford to attend regular classes for various preoccupations can pursue correspondence courses from various Open Universities and distance learning educational institutes in India.

ii. Higher Education Institutes in India:

Universities and its constituent colleges are the main institutes of higher education in India. There are at present two hundred and twenty seven (227) government-recognized Universities in India. Out of them twenty (20) are central universities, one hundred and nine (109) are deemed universities and eleven (11) are Open Universities and rest are state universities. Most of these universities in India have affiliated colleges where undergraduate courses are being taught. However Jawaharlal University is a remarkable exception to this rule. Apart from these higher education institutes there are several other private institutes in India that offer various professional courses in India. According to the Department of higher Education, government of India, there are 16,885 colleges, 99.54 lakh students and 4.57 lakh teachers in various higher education institutes in India.

iii. **Institutions Governing Higher Education:**

The institutions which are governing higher education in India are quite independent and multi-faceted. Among these institutions, University Grants Commission (UGC) is responsible for coordination, determination and maintenance of standards, release of grants.

Professional Councils are responsible for according recognition of courses, promotion of professional institutions and providing grants to undergraduate programmes and various awards. The statutory professional councils are:

a. All India Council for Technical Education (AICTE),
b. Distance Education Council (DEC)
c. Indian Council for Agriculture Research (ICAR),
d. Bar Council of India (BCI),
e. National Council for Teacher Education (NCTE)
f. Rehabilitation Council of India (RCI)
g. Medical Council of India (MCI),
h. Pharmacy Council of India (PCI)
i. Indian Nursing Council (INC)
j. Dentist Council of India (DCI)
k. Central Council of Homeopathy (CCH)
l. Central Council of Indian Medicine (CCIM)

Further, Central Government is responsible for major policy relating to higher education in the country. It provides grants to the UGC and establishes central universities in the country. The Central Government is also responsible for declaration of Educational Institutions as 'Deemed to be University' on the recommendation of the UGC.

Presently there are eighteen (18) Central Universities in the country. In pursuance of the Mizoram Accord, another Central University in the State of Mizoram is planned. There are ninety nine (99) Institutions which have been declared as Deemed to be Universities by the Govt. of India as per Section of the UGC Act, 1956. State Governments are responsible for establishment of State Universities and colleges, and provide plan grants for their development and non-plan grants for their maintenance.

The coordination and cooperation between the Union and the States is brought about in the field of education through the Central Advisory Board of Education (CABE).

Special Constitutional responsibility of the Central Government: Education is on the 'Concurrent list' subject to Entry sixty six (66) in the Union List of the Constitution of India. This gives exclusive Legislative Power to the Central Govt. for co-ordination and determination of standards in Institutions of higher education or research and scientific and technical institutions.

iv.   Academic Qualification Framework – Degree Structure

There are three principle levels of qualifications within the higher education system in the country. These are:

a. Bachelor / Undergraduate level
b. Master's / Post-graduate level
c. Doctoral / Pre-doctoral level

Diploma courses are also available at the undergraduate and postgraduate level. At the undergraduate level, it varies between one to three years duration, postgraduate diplomas are normally awarded after one year's study.

Bachelor's degree in arts, commerce and sciences is three years of education (after 12 years of school education). In some places there are honours and special courses available. These are not necessarily longer in duration but indicate greater depth of study. Bachelor degree in professional field of study in agriculture, dentistry, engineering, pharmacy, technology and veterinary medicine generally take four years, while architecture and medicine, it takes five and five and a half years respectively. There are other bachelor degrees in education, journalism and library sciences that are second degrees. Bachelor's degree in law can either be taken as an integrated degree lasting five years or three-year course as a second degree.

Master's degree is normally of two-year duration. It could be coursework based without thesis or research alone. Admission to postgraduate programmes in engineering and technology is done on the basis of Graduate Aptitude Test in Engineering or Combined Medical Test respectively.

A pre-doctoral programme - Master of Philosophy (M.Phil.) is taken after completion of the Master's Degree. This can either be completely research based or can include course work as well. Ph.D. is awarded two year after the M.Phil. or three years after the Master's degree. Students are expected to write a substantial thesis based on original research, generally takes longer.

v. System of Governance of Higher Education Institutions:

The Universities are of various kinds: with a single faculty, or multi-faculties; teaching or affiliating, or teaching cum affiliating, single campus or multiple campuses. Most of the Universities are affiliating universities, which prescribe to the affiliated colleges the course of study, conduct examinations and award degrees, while undergraduate level and to some extent postgraduate level the colleges affiliated to them impart instruction at graduate level. Many of the universities along with their affiliated colleges have grown rapidly to the extent of becoming unmanageable. Therefore, as per National Policy on Education, 1986, a scheme of autonomous colleges was promoted. In the autonomous colleges, whereas the degree continues to be awarded by the University, the name of the college is also included. The colleges develop and propose new courses of study to the university, for approval. They are also fully responsible for conduct of examination. There are at present one hundred and thirty eight (138) autonomous colleges in the country.

vi. Central Universities

In addition to the Universities, operating under the purview of University Grants Commission, there are eighteen (18) Universities which are functioning under the supervision of Govt. of India. By virtue of their functioning and needs created by them these institutions are covered under special provisions. Some of these provisions are:

- President of India is the Visitor of all Central Universities.

- President/Visitor nominates some members to the Executive Committee/Board of Management/Court/Selection Committees of the University as per the provisions made in the relevant University Act.

- Ministry provides secretariat service for appointment of Vice-Chancellor/Executive Committee Nominees/Court Nominees/Selection Committee Nominees etc. by the President.

There are eighteen (18) Central Universities under the purview of the MHRD which have been set up by Acts of Parliament. These universities are as follows:

i. University of Delhi, Delhi
ii. Jawaharlal Nehru University, New Delhi
iii. Jamia Millia Islamia, New Delhi
iv. Indira Gandhi National Open University, New Delhi
v. Banaras Hindu University, Varanisi
vi. Aligarh Muslim University, Aligarh
vii. Viswa Bharati, Santiniketan
viii. Hyderabad University, Hyderabad
ix. Pondicherry University, Pondicherry

x. North-Eastern Hill University, Shillong

xi. Assam University, Silchar

xii. Tezpur University, Tezpur

xiii. Nagaland University, Kohima

xiv. Babasaheb Bhimrao Ambedkar University, Lucknow

xv. Maulana Azad National Urdu University, Hyderabad

xvi. Mahatma Gandhi Antarasshtriya Hindi Vishwavidyalaya, Wardha

xvii. Mizoram University, Aizawal

xviii. University of Allahabad, Allahabad

These universities were established with a specific focus to cater to the needs of higher education in the country. A brief account on each university is as follows:

a. Jawaharlal Nehru University, New Delhi

Jawaharlal Nehru University came into existence in 1969 by an Act of Parliament. It is primarily concerned with Post-graduate Education and Research. The University has been identified by the University Grants Commission as one of the Universities in the Country with 'Potential for Excellence'. It has nine (9) schools consisting of twenty seven (27) centres of studies and four (4) special centres. The strength of its teaching and non-teaching staff is four hundred twenty (420) and one thousand two hundred and ninety seven (1297) respectively. Four thousand eight hundred and ninety (4890) students were on rolls.

b. University of Delhi

University of Delhi was established in February, 1922 as a unitary and residential university. It has fourteen (14) faculties, eighty two (82) teaching departments and seventy eight (78) colleges spread over national Capital Territory of Delhi. A new State University – Indraprashtha Vishwavidhlaya has come up in Delhi as an affiliating University.

c. Jamia Millia Islamia

Functioned as a Deemed University since 1962, and acquired the status of a Central University in December 1988 by an Act of Parliament, imparts education from nursery stage to Post-graduate and Doctorate levels. It has twenty nine (29) Departments, excluding the various Centres of Studies and Research, grouped under seven (7) Faculties, offering a total of one hundred and twenty one (121) courses at the undergraduate and postgraduate levels, in addition to Ph.D Programmes. It has on its rolls a total of 14,000 students, including ninety seven (97) foreign students from thirty eight (38) countries. The total strength of the teaching staff is six hundred and twelve (612) (including 120 for School Sector) and that of the non-teaching staff is nine hundred and ninety seven (997). Apart from providing training at Postgraduate level in Mass Communication, A.J. Kidwai Mass Communication Research Centre produces material on different educational aspects/subjects for the UGC's Indian National Satellite (INSAT) Programme.

The new initiatives taken in the areas of academics include setting up of the various new Centres, including (i) Centre for

Jawaharlal Nehru Studies, (ii) Centre for West Asian Studies, (iii) Centre for Dalit and Minorities Studies, (iv) Centre for Spanish and Latin American Studies and (v) Centre for Comparative Religion.

d. Indira Gandhi National Open University (IGNOU)

Established in 1985 by an Act of Parliament for introduction and promotion of Open University and distance education system in the Country. Major objectives include widening of access to higher education. During 2005 the University offered one hundred and one (101) programmes and the total number of students registered for various programmes reached over 3, 60,000. Its Students 'Support Services consist of forty-eight (48) Regional Centres, six Sub Regional Centres and one thousand and two hundred (1200) Study Centres. IGNOU programmes telecast on National Channel (DD-1) in the morning slot and Gyan Darshan, exclusive education channel. At present, the University has bouquet of six digital channels of Gyan Darshan, Under the **Gyan Vani FM** Radio initiative, the University has set up seventeen (17) Radio Stations. Today IGNOU has set up its study centres and has entered into collaboration with educational institutions in thirty five (35) countries. Distance Education Council (DEC) under IGNOU has the responsibility for coordination and maintenance of standards in open and distance education system in the country.

e. Aligarh Muslim University

Established in the year 1920, as a fully residential Central University, It has one hundred and two (102) Departments/Institutions/Centres/units grouped under twelve

(12) faculties. It also maintains four Hospitals, six Colleges (including Medical, Dental and Engineering Colleges), two Polytechnics and eight Schools. The university offers six diploma-level vocational courses exclusively for women. Nineteen thousand seven hundred and three (19,703) students (excluding its secondary schools' strength) drawn from twenty five (25) States of the country are on its rolls. Strength of the teaching staff is one thousand four hundred and fifty seven (1,457) and that of non-teaching staff is five thousand eight hundred and ninety nine (5,899).

f.   Banaras Hindu University, Varanasi:

Banaras Hindu University came into existence in the year 1916 as a teaching and residential University. It consists of three Institutions - Institute of Medical Sciences, Institute of Technology and Institute of Agricultural Sciences. It has faculties with one hundred and twenty one (121) academic departments and four (4) Inter-disciplinary schools. It maintains a constituent Mahila Mahavidyalaya and three School level institutions. A thousand (1000) bedded Modern/Ayurvedic Medicine Hospital. It has fourteen thousand eight hundred and twelve (14, 812) students in the rolls. Teaching staff are one thousand one hundred and sixty two (1,162) and non-teaching seven thousand eighty-eight (7,088).

g.   University of Hyderabad:

Established in 1974 as one of the premier Central Universities, which has been identified as a 'University with Potential for Excellence' by the University Grants Commission? The University was established for post graduate teaching and research and is twenty two (22) KM from the city of Hyderabad on the Old Hyderabad - Bombay road. It has a City Campus - The Golden Threshold- the residence of the late Smt. Sarojini Naidu. The University has Eight Schools of Studies and a Centre for Distance Education offering post-graduate diploma in five disciplines. The student strength of the University during the year 2003-04 was two thousand four hundred and seventy seven (2477) out of which fifty one (51) per cent are research students pursuing Ph.D and M.Phil/M.Tech Programmes in various disciplines. University of Hyderabad Home page and web site is http://www.uohyd.ernet.in.

h. Visva Bharati University

An educational institution founded by late Gurudev Rabindranath Tagore in 1921 was declared in 1951, by an Act of Parliament, as an institution of national importance to provide for its functioning as a unitary, teaching and residential University. Its jurisdiction is restricted to the area known as Santiniketan in the district Birbhum, West Bengal. It imparts education from the Primary School level to Post-graduate and Doctorate levels. It has twelve institutes - eight at Santiniketan, three at Sriniketan and one at Kolkata. It has on its rolls a total of six thousand two hundred and twenty seven (6,227) students, including its Schools' strength. The total strength of teaching and non-teaching staff is five hundred fifty (550) and one thousand four hundred (1,400) respectively. Admission is on the basis of merit adjudged through Admission Test.

Palli Samgathana Vibhaga, originally established by Gurudev Tagore in Sriniketan in 1922 with the primary objective to bring about regeneration of village life through self-reliance in the village around Santiniketan and Sriniketan, is now one of the Institutes of the University with four Departments - Rural Extension Centre, Silpa Sadana (a pioneering institute in India in developing cottage and small scale industries), Department of Social Work and Palli Charcha Kendra (offering a Ph.D. programme and two PG level courses - in Anthropology and Rural Development - apart from conducting research on various aspects of the life of the rural people).

i. **North Eastern Hill University:**

Established in 1973 at Shillong by an Act of Parliament with the object to disseminate and advance knowledge by providing instructional and research facilities in various branches of learning and also to pay special attention to the improvement of the social and economic conditions and welfare of the people of the hilly areas . The University has its jurisdiction over the State of Meghalaya and has a Campus at Tura. The North Eastern Hill University (NEHU) presently has twenty four (24) academic departments and four centres of studies under six schools. There are fifty eight (58) under graduate colleges and eight (8) professional colleges affiliated to the University. The number of students studying for their Master's Degrees and research students working for their M.Phil and Ph.D degrees is close to one thousand seven hundred (1700). The faculty strength of the University is over three hundred (300). The undergraduate colleges affiliated to the University enroll about nineteen thousand five hundred (19,500) students and the e-mail address is khathing@nehu.ac.in.

k. Pondicherry University, Pondicherry:

It is established by an Act of Parliament in the year 1985 as a teaching-cum-affiliating university with its jurisdiction over the Union Territories of Pondicherry and Andaman and Nicobar Islands. It has eight (8) Schools, twenty four (24) Departments and nine (9) Centres, and it offers Certificate Course in one discipline, Post-graduate programmes in twenty six (26) disciplines, M.Tech. in one discipline, M. Phil programme in twenty three (23) disciplines, Ph.D. programme in twenty six (26) disciplines and PG Diploma programme in six (6) disciplines. The University has forty one (41) affiliated colleges of which twenty seven (27) are located in Pondicherry, five (5) in Karaikal, two (2) in Mahe, two (2) in Yanam and five (5) in Andaman and Nicobar Islands. Total students strength in these institutions is twenty one thousand and thirty four (21034). Students' enrolment in the University is one thousand five hundred and thirty eight (1538), out of which three hundred and twelve (312) students belong to SC/ST and women students are five hundred thirty six (536). University has faculty strength of one hundred thirty six (136) teachers and five hundred twenty three (523) non-teaching staff. Thirty five (35) Research Scholars have registered for the Ph.D. programme. Sixty eight (68) sponsored Research Projects of topical relevance are in progress.

l. Nagaland University

It was established in 1994 as a teaching-cum-affiliating University with Head quarters at Lumami, Nagaland with the objective to disseminate and advance knowledge by providing instructional and research facilities and to make provision for various integrated courses, innovations in teaching-learning process and to pay special attention to the improvement of social and economic conditions and welfare of the people of Nagaland, their intellectual, academic and cultural development. The University has jurisdiction over the whole of the State of Nagaland and has twenty five (25) Departments and four (4) Schools of studies with forty seven (47) colleges affiliated to it. The University has campuses in Kohima, Lumami and Medziphema (School of Agricultural Sciences and Rural Development- SASRD). Faculty strength of the University is one hundred and nine (109) and Students enrollment is eighteen thousand and seventy eight (18,078).

m. Tezpur University

A non-affiliating unitary Central University set up in 1994 under an Act of Parliament. Its aim is to offer employment-oriented and inter-disciplinary courses, mostly at post-graduate level. The University has twelve (12) Departments under four (4) Schools of studies and six (6) centres of Studies with seventy nine (79) faculty members and Students enrolment of six hundred forty eight (648).

n. Assam University

Established as a teaching-cum-affiliating University on DT: 21.1.1994 with the objective to disseminate and advance knowledge by providing instructional and research facilities in various branches, to take appropriate measures for promotion of inter-disciplinary studies and research in the University and also to pay special attention to the improvement of the social and economic conditions and welfare of the people of the State. The University has its jurisdiction over the districts of Cachar, Karimganj, Karbi, Anglong and Hailakandi in the State of Assam and has fifty three (53) affiliated colleges with twenty four (24) Departments under eight (8) Schools of studies and three (3) Centres of studies. Total number of teachers is one hundred and one (101) and students on the rolls, including those in affiliated colleges, are eighteen thousand and seventy eight (18,078).

o. Mahatma Gandhi Antarrashtriya Hindi Viswavidyalaya, Wardha.

Mahatma Gandhi Antarrashtriya Hindi Vishwavidyalaya was set up during the year 1997 by an Act of Parliament with its headquarter at Wardha to promote and develop Hindi language and literature in general. The University started functioning as a camp office in a rented house at New Delhi. The University was shifted from its camp office to it's headquarter at Wardha in December, 2003. The teaching activities of the University started with two courses in the academic session 2002-03. Presently the University is running five MA courses and two M.Phil courses.

p. Babasaheb Bhimrao Ambedkar University, Lucknow

Established as a State University in 1994 at Lucknow, it was notified as a Central University on 10th January 1996. It aims to provide instructional and research facilities in new and frontier areas of learning. It has 5 Schools comprising 8 Departments viz. (1) School of Ambedkar Studies, (2) School of Biosciences and Bio-technology, (3) School of Environmental Sciences, (4) School of

Information Science and Technology, (5) School of Legal Studies. Total enrolment of students is two hundred and fifty five (255), including 35 Ph.D. scholars, during the year 2004-05 out of which ninety five (95) (37.25 percent) belong to the SC/ST category.

q.   Mizoram University

The Mizoram University, with its headquarters at Aizawl, was established as a teaching and affiliating university with effect from the 2nd July, 2001.The academic activities of the University are presently carried out through its sixteen teaching departments and one constituent college. The total number of students enrolled in these departments and the constituent college is one thousand one hundred and eighty seven (1,187) and the teaching and non-teaching staff during the year 2004-2005 was one hundred twenty four (124) and two hundred and thirty four (234) respectively. Besides, the University has twenty eight (28) affiliated colleges located in the State of Mizoram. The number of students studying in these affiliated colleges is five thousand five hundred and seventy nine (5,579).

r.   Maulana Azad National Urdu University, Hyderabad

The aim of Maulana Azad National Urdu University is to promote and develop Urdu language and to impart vocational and technical education in Urdu medium through conventional and distance education system. The University was established in 1997 by an Act of Parliament. Its Administrative head quarters have been set up at Hyderabad and have five Regional Centres at Delhi, Patna, Bangalore, Bhopal and Dharbhanga. The University has so far established eighty four (84) Study Centres spread over in fourteen (14) States of the Country.

vii. **National Assessment and Accreditation Council (NAAC)**

National Assessment and Accreditation Council (NAAC) is an autonomous institutions established by the University Grants Commission in 1994 NAAC's responsibility is to assess and accredit institutions of higher education that volunteer for the process, based on certain prescribed criteria. NAAC's process of assessment and accreditation involves the preparation of a self -study report by the institution, its validation by the peers and final decision by the Council. One hundred twenty two (122) universities and two thousand four hundred eighty six (2486) colleges/ institutions have been accredited by NAAC so far.

viii. **Open University System**

The advances in information and communication technology provide great opportunities to enhance teaching and learning in higher education by both on-campus and distance education. Even disabled students who are denied access to traditional institutions, and all those who require updating of their knowledge and life-long education can now be benefited by the modern facilities of communication. They also provide increased access to information sources and facilitate communication among researchers and teachers and the building of networks of institutions and scholars.

Through the open universities and distance learning initiatives, mechanisms are in place to upgrade skills at regular intervals and develop new competencies. People's needs of lifelong learning are constantly expanding. Higher education institutions are offering learning opportunities to satisfy these diverse demands. Ready access and flexibility are the hallmarks of these initiatives.

The Open University System was initiated in the country to augment opportunities for higher education as an instrument of democratising education and also to make it a lifelong process. The first Open University in the country was established by the state government of Andhra Pradesh in 1982. In 1985, the central government established the Indira Gandhi National Open University (IGNOU).

ix. Distance Education Council (DEC)

The IGNOU is also a national level apex body for distance education. Distance Education Council has been established as a statutory authority under the IGNOU Act. The DEC is responsible for promotion, coordination and maintenance of standards of open and distance education system in the country.

The apex body's role envisages the establishment and development of an Open University Network by sharing the intellectual and physical resources within the distance education system among different institutions and enriching the system by extending its outreach, on the one hand, and ensuring the quality standards of its programmes of education and training, on the other. In discharging its responsibility, the Distance Education Council also provides development funding to open universities and distance education institutions from the funds placed at its disposal by the Central Government. The DEC has been supporting all the

State Open Universities (SOUs) since eighth plan and Distance Education Institutions (DEIs) of conventional universities since ninth plan. Presently, there is one National Open University (IGNOU), eleven (SOUs) and a number of DEIs in different states.

x. Open University Network

The Distance Education Council (DEC) has taken several initiatives to develop the Open University Network. The programmes developed and produced by IGNOU are extensively used by the State Open Universities in the country. Efforts have also been made to evolve common standards for the products as well as processes (programme structure, credits, examination, grading, etc.) to facilitate student mobility across programmes as well as institutions through systems of credit transfer. Steps have also been initiated of credit transfer. Steps have also been initiated to frame norms and standards for the design, development and delivery of programmes in specific fields and to ensure their quality.

xi. Important Institutions of Higher Education: A Brief Review

Having understood the various facets of higher education in India, an attempt was made to focus on functioning of a few of the important institutions which govern higher education in India.

a. University Grants Commission (UGC)

The Government established university Grants Commission (UGC) by an Act of Parliament in 1956. It discharges the Constitutional mandate of coordination, determination, and maintenance of standards of teaching, examination and research in the field of University and Higher Education. UGC serves as a vital link between the Union and State Governments and the institutions of higher learning. It monitors developments in the field of collegiate and university education; disburses grants to the universities and colleges; advises Central and State Governments on the measures necessary for the improvement of university education; and frames regulations such as those on the minimum standards of instruction

The Commission comprises the Chairperson, Vice-Chairperson and ten other members appointed by the Central Government. The Chairperson is selected from among persons who are not officers of the Central Government or any State Government. Of the ten members, two are from amongst the officers of the Central Government to represent it. Not less than four, selected from among persons who are, at the time they are selected, shall be a teacher in the Universities. Others are selected from among eminent educationists, academics and experts in various fields.

Chairperson is appointed for a term of five (5) years or until the age of sixty five (65) years, whichever is earlier. Vice-Chairperson is appointed for a term of three (3) years or until the age of sixty five (65) years, whichever is earlier. The other members are appointed for a term of three (3) years. The Chairperson, Vice-Chairperson and members can be appointed for a maximum of two terms.

UGC has no funds of its own. It receives both Plan and Non-Plan grants from the Central Government to carry out the responsibilities assigned to it by law. It allocates and disburses full maintenance and development grants to all Central Universities, Colleges affiliated to Delhi and Banaras Hindu Universities and some of the institutions accorded the status of 'Deemed to be Universities'. State Universities, Colleges and other institutions of higher education, receive support only from the Plan grant for development schemes. Besides, it provides financial assistance to Universities and colleges under various schemes/programmes for promoting relevance, quality and excellence as also promoting the role of social change by the Universities.

b.   All India Council for Technical Education (AICTE)

The All India Council for Technical Education (AICTE), is the statutory body established for proper planning and co-ordinated development of the technical education system in India. It was established in November, 1945. Currently, there are one thousand three hundred fourty six (1,346) engineering colleges in India approved by the All India Council of Technical Education with a seat capacity of four lakh fourty thousand (4,40,000).

All the major Engineering, Pharmacy and MBA colleges are affiliated with AICTE. The prominent exceptions are B. Tech. courses from Indian Institute of Technology (IITs), ICFAI and BITS, Pilani. The AICTE has its Headquarters in Indira Gandhi Sports Complex, Indraprastha Estate, New Delhi, which has the offices of the Chairman, Vice-Chairman and the Member Secretary

AICTE is vested with the statutory authority for planning, formulation and maintenance of norms and standards, quality assurance through school accreditation, funding in priority areas, monitoring and evaluation, maintaining parity of certification and awards and ensuring coordinated and integrated development and management of technical education in the country as part of the AICTE Act No. 52 of 1987.

The AICTE Act, stated verbatim reads:

To provide for establishment of an All India council for Technical Education with a view to the proper planning and co-ordinated development of the technical education system throughout the country, the promotion of qualitative improvement of such education in relation to planned quantitative growth and the regulation and proper maintenance of norms and standards in the technical education system and for matters connected therewith.

In order to improve upon the present technical education system, the current objectives are to modify the engineering curriculum as follows:

1. Greater emphasis on design oriented teaching, teaching of design methodologies, problem solving approach.

2. Greater exposure to industrial and manufacturing processes.

3. Exclusion of outmoded technologies and inclusion of the new appropriate and emerging technologies.

4. Greater input of management education and professional communication skills.

c. Indian Council of Agricultural Research

The Indian Council of Agricultural Research (ICAR), New Delhi, India is the apex body in Agriculture and related allied fields, including research and education. The Union Minister of Agriculture is the President of the ICAR. Its principal officer is the Director-General who is also the Secretary to the Government of India in the Department of Agricultural Research and Education (DARE).

ICAR has two bodies.

1. The General Body, the supreme authority of the ICAR, is headed by the Minister of Agriculture, Government of India

2. The Governing Body which is the chief executive and decision making authority of the ICAR. It is headed by the Director-General.

As of July, 2006 it has developed a vaccine against bird flu. The vaccine was developed at the High Security Animal Disease Laboratory, Bhopal, the only facility in the country to conduct tests for the H5N1 variant of bird flu. It was entrusted with the task of developing a vaccine by the ICAR after the Avian Influenza outbreak in February. The ICAR provided an amount Rs. 8 crore for the purpose.

xii. Higher Education in India: A Few Milestones

After elucidating on the institutions responsible for promotion of higher education in India, the following are the details of a few milestones in terms of higher education in India.

Year	Event
1948-49	University Education Commission constituted;
1951	First Indian Institute of Technology (IIT) established at Kharagpur
1956	University Grants Commission (UGC) established by Act of Parliament
	• Indian Institute of Technology (Kharagpur) Act passed by Parliament
	• Pandit Jawaharlal Nehru delivers the first convocation address at the first IIT (Kharagpur)
1958	Second IIT established at Mumbai
1959	Third and Fourth IITs established at Kanpur and Chennai, respectively
1961	Institutes of Technology Act passed by Parliament to provide a common legal framework for all IITs
	First two Indian Institutes of Management (IIMs) set up at Ahmedabad and Kolkata
1963	Fifth IIT established at Delhi
1964-66	Education Commission constituted; submits Report
1968	First National Policy on Education (NPE) adopted, in the light of the recommendations of the Education Commission
1963	Third IIM established at Banglore
1976	Constitution amended to change "Education" from being a "State" subject to a "Concurrent" subject
1984	Fourth IIM established at Lucknow
1985	Indira Gandhi National Open University (IGNOU) established by an Act of Parliament
1986	New National Policy on Education (NPE) adopted

1987-88	All India Council of Technical Education (AICTE) vested with statutory status by an Act of Parliament
1992	NPE, 1986, revised, based on a review by the Acharya Ramamurti Committee
1994	National Assessment and Accreditation Council (NAAC) established by UGC (with headquarters at Bangalore) to assess and accredit institutions of higher education
	National Board of Accreditation (NAB) established by AICTE to periodically evaluate technical institutions and programmes
	Sixth IIT established at Guwahati
1996	Fifth IIM established at Kozhikode
1998	Sixth IIM established at Indore
2001	University of Roorkee converted into (the seventh) IIT
2003	seventeen (17) Regional Colleges of Engineering converted into National Institutes of Technology, fully funded by the Central Government
2006	Two Indian Institutes of Science Education & Research (IISERs) established at Kolkata and Pune, respectively

Challenges before Indian Education

Having reviewed the pattern of higher education in India and the institutions responsible for the same, a revision was attempted to understand the challenges existing in the present day circumstances. These challenges spread over various sectors of higher education. Some of the main challenges before the Education System in India pertain to:

- Access,

- Participation & Equity,

- Quality,
- Relevance,
- Management, and
- Resources.

i. Access:

While availability of elementary schools within a reasonable distance from habitations is now fairly universal, same cannot yet be said in regard to Secondary Schools and Colleges. Pockets still exist in many remote parts of the country where the nearest Secondary School or College is much too far for everyone to be able to attend. Besides the physical availability of institutions, other barriers to access – e.g. socio-economic, linguistic-academic, physical barriers for the disabled, etc. – also need to be removed.

ii. Participation and Equity:

Gross Enrolment Ratios for the elementary, secondary and tertiary stages of education in 2003-04 were 85%, 39% and 9%, respectively. These participation rates are undoubtedly low, and need to be raised very substantially, for India to become a knowledge society / economy.

A linked challenge is one of equity. Participation rates in Education are poor largely because students from disadvantaged groups continue to find it difficult to pursue it. Even when they manage to participate, students suffering from disadvantages of gender, socio-economic status, physical disability, etc. tend to have access to education of considerably lower quality than the others, while the education system needs to provide them access to the best possible education so that they are able to catch up with the rest.

iii. Quality:

The challenge of quality in Indian education has many dimensions, e.g. providing adequate physical facilities and infrastructure, Making available adequate teachers of requisite quality, Effectiveness of teaching-learning processes, Attainment levels of students, etc. Besides the need to improve quality of our educational institutions in general, it is also imperative that an increasing number of them attain world-class standards and are internationally recognized for their quality.

iv. Relevance

Education in India needs to be more skill-oriented – both in terms of life-skills as well as livelihood skills. In sheer numerical terms, India has the manpower to substantially meet the needs of a world hungry for skilled workers, provided its education system can convert those numbers into a skilled work-force with the needed diversity of skills.

v. Management

Management of Indian education needs to build in greater decentralization, accountability, and professionalism, so that it is able to deliver good quality education to all, and ensure optimal utilization of available resources.

vi. Resources

India's stated national policy - ever since 1968 - has been to raise public expenditure on Education to the level of six percent (6%) of GDP. On the other hand, in 2004-05, outlay of Central and State Governments for Education amounted to about three and half (3.5%) of GDP. Thus, the gap in allocations for Education is still substantial, and needs to be urgently bridged.

In the next chapter the aspects of research methodology and the ways and means and the tools utilized to handle data, interpretational aspects and a detailed description of study area were taken up.

"**By learning you will teach; by teaching you will learn.**"

- Latin **proverb**

Chapter – 4

Abstract

In this chapter statement of the problem, its precedent conditions, data collection, methods, treatment, major objectives, research tools utilized and the allied processes for were out lined. Along with this research design the limitations of the research study and research design how it is designed and executed is given. To conclude the actual process of how the investigation was carried out from introduction, chapterization and up to summary and conclusions is detailed in this chapter.

RESEARCH METHODOLOGY

"No one can become really educated without having pursued some study in which he took no interest. For it is part of education to interest ourselves in subjects for which we have no aptitude."

_____ T. S. Eliot

I. Introduction

The need for education evolved and unfolded itself with the realisation of the fact that it is the key to the future and progress of mankind (Roy, 1967). This in background, for a country like India, with all its manifestations of socio-economic and technical backwardness has the only alternative way to implement education related programmes on a priority basis. The very realisation resulted in launching emphasis on opening the doors for private sector to spread education at all levels i.e. primary to higher education.

Every individual is born with a collection of abilities and talents. Education in its many form, has the potential to help fulfill and help them in order to achieve social and economic development. It has become a common practice for national development to be linked to education, with development frequently measured in terms of education (Wagner, 1990).

There is general agreement, among scholars that one of the fundamental breakthroughs in the emergence of civilization was the invention of writing as a means of communication. With a written word it has become possible for historical events to be accurately recorded, and for knowledge to be more widely and quickly disseminated among several generations. In writing about importance of education, Goody and Watt (1977) point out that not only was trade, commerce and economic sector of the society radically altered, but also the nature of human interaction transformed.

In brief, an educated individual has greater powers of communication, critical consciousness and control over his or her environment. The mobilization of human potential for social and collective action in ancient Egypt, Babylonia and Greece and even with partially literate populations dramatically surpassed that of pre-literate tribes or nomadic groups. Education is also a basic human right which expands personal choice, over one's own environment, and allows for collective action not control otherwise possible. Much of the concern today in under-developed and developed countries about education stems, at least in part, from this consideration. (Ingemer and Saha, 1989).

Changes in the society are coming thick and fast. The changes are explicit in the sectors of economy and technology. They call for new shape of schools, new learner profile, teacher profile, and administrator's profile. More attention has to be paid now to excellence, quality and efficiency so that peace and harmony in the society can be maintained. Acceptance and appreciation for diversity and pluralism is inevitable. The future agenda for education will be to empower individuals, assure high quality of life and pave way to learning society

Research Methodology

Two of the finest statements of educational vision in India must be recalled while attempting a discussion on education. The first of these was made by Gandhiji in 1931 in London in the context of universalising education in India.

'India lives in its villages. It is there that our producers live, voters live, the poor and illiterate live. It is the villages that hold the key to the country's problems. So vision of future India can be greater than to rebuild its half a million villages. The irony is that in terms of the teaming millions inhabiting these villages our developments, our democracy, and our education have all become irrelevant. But once we decide to approach them in the right spirit they are bound to respond, and rise to end their suffering. It may be that in the first phase selected homogenous SC/ST and other backward villages may have to be taken up. In case whole villages do not come forward in the beginning, then mutual-aid teams may have to be formed. Naturally in the whole process of rebuilding villages education will have the most vital part to play, because it alone can prepare people's mind to receive new ideas, and accept new tools, new relationships, and new forms of organization'.

The second vision statement is contained in Article 45 of the Constitution of India: The state shall endeavour to provide within a period of ten years from the commencement of this Constitution, for free and compulsory education for all children until they complete the age of fourteen years. As institutions of higher learning involved in the education of current and future managers we are voluntarily committed to engaging in a continuous process of improvement of the following Principles and their application, reporting on progress to all our stakeholders and exchanging effective practices with other academic institutions:

i. Principle 1

Purpose: develop the capabilities of students to be future generators of sustainable value for business and society at large and to work for an inclusive and sustainable global economy.

ii. Principle 2

Values: incorporate into our academic activities and curricula the values of global social responsibility as portrayed in international initiatives such as the United Nations Global Compact.

iii. Principle 3

Method: create educational frameworks, materials, processes and environments that enable effective learning experiences for responsible leadership.

iv. Principle 4

Research: engage in conceptual and empirical research that advances our understanding about the role, dynamics, and impact of social and economic initiatives in the creation of sustainable social, environmental and economic value.

v. Principle 5

Partnership: interact with like-minded people to extend our knowledge of their challenges in meeting social and environmental responsibilities and to explore jointly effective approaches to meeting these challenges.

vi. Principle 6

Dialogue: facilitate and support dialogue and debate among educators, business, government, consumers, media, civil society organizations and other interested groups and stakeholders on critical issues related to global social responsibility and sustainability. Thus, the ethical practices and management issues in education matters the most in shaping the future of the country.

Higher education in India is gasping for breath, at a time when India is aiming to be an important player in the emerging knowledge economy. With about 300 universities and deemed universities, over 15,000 colleges and hundreds of national and regional research institutes, Indian higher education and research sector is the third largest in the world, in terms of the number of students it caters to. However, not a single Indian university finds even a mention in a recent international ranking of the top 200 universities of the world, except an IIT ranked at 41, whereas there were three universities each from China, Hong Kong and South Korea and one from Taiwan (N. Raghuram, 2006).

While many reasons can be cited for this situation, they all boil down to decades of feudally managed, colonially modelled institutions run with inadequate funding and excessive political interference. Only about ten per cent (10%) of the total student population enters higher education in India, as compared to over fifteen per cent (15%) in China and fifty per cent (50%) in the major industrialised countries. Higher education is largely funded by the state and central governments so far, but the situation is changing fast. Barring a few newly established private universities, the government funds most of the universities, whereas at the college level, the balance is increasingly being reversed.

The experience over the last few decades has clearly shown that unlike school education, privatisation has not led to any major improvements in the standards of higher education and professional education. Yet, in the run up to the economic reforms in 1991, the IMF, World Bank and the countries that control them have been crying hoarse over the alleged pampering of higher education in India at the cost of school education. The fact of the matter was that school education was already privatised to the extent that government schools became an option only to those who cannot afford private schools mushrooming in every street corner, even in small towns and villages. On the other hand, in higher education and professional courses, relatively better quality teaching and infrastructure has been available only in government colleges and universities, while private institutions of higher education in India capitalised on fashionable courses with minimum infrastructure.

With the result, the last decade has witnessed many sweeping changes in higher and professional education: For example, thousands of private colleges and institutes offering IT courses appeared all across the country by the late 1990s and disappeared in less than a decade, with devastating consequences for the students and teachers who depended on them for their careers. This situation is now repeating itself in management, biotechnology, bioinformatics and other emerging areas. No one asked any questions about opening or closing such institutions, or bothered about whether there were qualified teachers at all, much less worry about teacher-student ratio, floor area ratio, class rooms, labs, libraries etc. All these regulations that existed at one time (though not always enforced strictly as long as there were bribes to collect) have now been deregulated or softened under the self-financing scheme of higher and professional education adopted by the UGC in the ninth five-year plan and enthusiastically followed by the central and state governments (N. Raghuram, 2006).

The economics of imparting higher education are such that, barring a few courses in arts and humanities, imparting quality education in science, technology, engineering, medicine etc. requires huge investments in infrastructure, all of which cannot be recovered through student fees, without making higher education inaccessible to a large section of students. Unlike many better-known private educational institutions in Western countries that operate in the charity mode with tuition waivers and fellowships (which is one reason why our students go there), most private colleges and universities in India are pursuing a profit motive. This is the basic reason for charging huge tuition fees, apart from forced donations, capitation fees and other charges. Despite huge public discontent, media interventions and many court cases, the governments have not been able to regulate the fee structure and donations in these institutions. Even the courts have only played with the terms such as payment seats, management quotas etc., without addressing the basic issue of fee structure.

It is not only students but also teachers who are at the receiving end of the ongoing transformation in higher education. The nation today witnesses the declining popularity of teaching as a profession, not only among the students that we produce, but also among parents, scientists, society and the government. The teaching profession today attracts only those who have missed all other "better" opportunities in life, and is increasingly mired in bureaucratic controls and anti-education concepts such as "hours" of teaching "load", "paid-by-the-hour", "contractual" teachers etc. With privatisation reducing education to a commodity, teachers are reduced to tutors and teaching is reduced to coaching. The consumerist boom and the growing salary differentials between teachers and other professionals and the value systems of the emerging free market economy have made teaching one of the least attractive professions that demands more work for less pay. Yet, the society expects teachers not only to be inspired but also to do an inspiring job!

II. Statement of the Problem

Though the issue of education spread has gained momentum, especially in the echelons of higher education, it was often reported in the academic circles as well as in the media that the outcome or products of the educational institutions are not conducive to the demands of the society. In other words, the education system is not in tune with the social and economic needs of the society and, thus, it is resulting in frictions in the society which would be quite difficult to deal with if left unattended to.

Further, it was often reported that ethical practices were pushed to a corner while managing educational institutions. The principal motive for the organisers of educational institutions was mere earning profits rather than producing a sound individual having all the ingredients that a society demands. Business has become the strong motive behind spreading educational institutions rather than focusing on ethical practices and better management aspects, is what the general criticism on the present day education system in the country.

Further, the three principal stakeholders of education i.e. Students, Parents and Educational Institutions, appears to be having different views and in the process what one can witness is sacrifices on the part of outcome of present day education. The ethical practices have always play an important role in providing appropriate education to the society so as to strengthen the society to meet the future requirements which is one of the bastion of education system.

However, research investigations into educational institutions and related stakeholders are quite few to understand the status of ethical practices and management process in the educational institutions. Especially, the roles and responsibilities of different stakeholders to ensure and protect ethical practices in education system were less emphasized in the ongoing research activities. Since, managing the education system in an appropriate manner is quite essential to ensure producing responsible citizens to take care of the future needs of the country.

Research Methodology

The understanding on issues related to ethical practices as well as management processes always provide an opportunity to understand the current educational practices so as to elucidate oneself with the contemporary situation as well as identifying the lacunae and corrective measures.

The above in view, the present study "THE ROLE OF ETHICS AND ENVIRONMENT IN EDUCATIONAL MANAGEMENT IN HIGHER EDUCATION: A CASE STUDY IN VISAKHAPATNAM" is taken up.

III. Research Objectives

The objectives of the research are furnished below:

1. To understand the social and economic profile of students participating in higher education.

2. To assess the perception and preferences of various students in higher education in their institutions about the ethical and management practices followed.

3. To assess organisational and managerial issues and problems encountered by promoters of higher education.

4. To assess the understanding and analyse opinion of students on the moral values protected and promoted by the educational institutions among stakeholders in higher education.

5. Suggest measures for better ethical and management practices in higher education.

IV Hypotheses

The research investigation is empirical in nature and hence no hypotheses were proposed.

V. Research Methodology

i. Study Area

The study is conducted in Visakhapatnam district in **Andhra Pradesh**, India. The district is specifically selected since Visakhapatnam is the seat of higher education for quite a long time. Institutions of Higher Education are operating with Visakhapatnam as a hub because of its situational, social, economic and industrial advantage. Studying the ethical and management practices in higher education in Visakhapatnam would certainly provide better insight and hence the selection is made.

ii. Selection of Sample (Respondents)

The respondents for the study consist of students who are pursuing higher education in the study area i.e. Visakhapatnam district in various educational institutions. Keeping the resources available with the researcher, a sample size of 300 students, 100 from engineering stream, 100 from commerce and management and 100 from arts and social science, were selected for the purpose of the study. The sampling was done on the basis of random sampling method to avoid prejudice in selection of respondents for the study.

Research Methodology

iii. Data Collection

The data was collected personally by the researcher in three months time in the field. The researcher personally visited the colleges offering the courses of bachelor of engineering, Master of Business Administration and Bachelor of Arts and commerce, established in Visakhapatnam district. Some times the researcher faced difficulty in contacting the students due to holidays and their examination schedules. Primary data was collected from the students through a questionnaire, and principal, correspondent, and faculty were interviewed and the notes were taken.

Secondary data was collected from books, journals, newspapers, reports and other printed material. Data collection was done through simple frequency tables and interpretations were made according to the objectives selected for the study by using the percentages, standard deviation, and t 'and 'F' tests where ever necessary.

iv. Data collection instruments

For the purpose of data collection, a questionnaire was exclusively designed and utilised. This questionnaire was prepared keeping in view the objectives identified for the research study. Accordingly, the questionnaire for students, of higher education was drafted by using the Likiret five point attitude scales. The questionnaire was pre tested before going for final data collection to avoid any ambiguities/deficiencies.

More over, the investigator had personal interview and had discussion with the principal, correspondent (Management

Representative), faculty, regarding practices adopted to inculcate ethical values with students and the environment provided in teacher-learning process. The details of the questionnaire are provided in annexure.

V.   Chapterisation

The report is presented in six chapters as mentioned below:

i.   Chapter I: Introduction
ii. Chapter II: Review of Literature
iii. Chapter III: Higher Education in India
iv. Chapter IV: Research Methodology
v.   Chapter V: Analysis of Data and Presentation of Report
vi. Chapter VI: Summary and Conclusions

Limitations of the Study:

1. The study is limited to students pursuing higher education in institutions of Visakhapatnam district.

2. No control group method was used.

3. As the appropriate standardized questionnaire will not be readily available, the researcher constructed the questionnaire by using Likert's five point attitude measurement scales. After pre-test the questionnaire was canvassed in the field.

Research Methodology

STUDY AREA: A DESCRIPTION

The study is carried out in Visakhapatnam district and this in view a comprehensive view on the profile of the district is presented in this chapter. Visakhapatnam is rich in mineral sources and in fact the industrial hub of Andhra Pradesh. The Visakhapatnam city, the district headquarters, is the seat of Andhra University, one of the oldest universities in southern India. The chapter is presented in two sections. The first section is devoted to profile of the district and the second section is devoted to profile of Andhra University under whose aegis the higher education is administered in the district.

Section I: Profile of Visakhapatnam District

Visakhapatnam District is one of the North Eastern Coastal districts of Andhra Pradesh and it lies between 17º - 15' and 18º-32' Northern latitude and 18º - 54' and 83º - 30' in Eastern longitude. It is bounded on the North partly by the Orissa State and partly by Vizianagaram District, on the South by East Godavari District, on the West by Orissa State and on the East by Bay of Bengal.

i. Historical Aspects

Inscriptions indicate that the District was originally a part of Kalinga Kingdom subsequently conquered by the Eastern Chalukyas in the 7th Century, A.D. who ruled over it with their Head Quarters at Vengi. This District was also under the occupation of various rulers such as the Reddy Rajahs of Kondaveedu, the Gajapathis of Orissa, the Nawabs of Golkonda and the Moghal Emperor Aurangazeb through a Subedar. This

LOCATION OF STUDY AREA MAP

INDIA

ANDHRA PRADESH

VISAKHAPATNAM

Andhra University, Visakhapatnam

territory passed on to French occupation in view of succession dispute among Andhra Kings and finally it came under the British reign. There were no geographical graftings till 1936 in which year, consequent on the formation of Orissa State the Taluks namely Bissiom, Cuttack, Jayapore, Koraput, Malkanagiri, Naurangapur, Pottangi and Ryagada in their entirety and parts of Gunpur, Paduva and Parvathipur Taluks were transferred to Orissa State.

The Visakhapatnam District was reconstituted with the remaining area and residuary portions of Ganjam District namely Sompeta, Tekkali and Srikakulam Taluks in entirety and portion of Parlakimidi, Ichchapuram, Berahmpur retained in Madras presidency. With the passage of time, the reconstituted District was found administratively unwieldy and therefore it was bifurcated into Srikakulam and Visakhapatnam districts in the year 1950. The residuary district of Visakhapatnam was further bifurcated and the Taluks of Vizianagaram, Gajapathinagaram, Srungavarapukota and portion of Bheemunipatnam Taluk were transferred to the newly created Vizianagaram District in the year 1979.

Coming to etymology of the name Visakhapatnam, tradition has it that some centuries ago a King of Andhra Dynasty encamped on the site of the present Head Quarters Town of Visakhapatnam on his piligrimage to Banaras and being pleased with the place, had built a shrine in honour of his family deity called Visakeswara to the South of the Lawsons Bay from which the district has derived its name as Visakheswarapuram which subsequently changed to Visakhapatnam. The encroachment of waves and currents of the sea supposed to have swept away the shrine into off shore area.

ii. Physical Features

The District presents two distinct Geographic divisions. The strip of the land along the coast and the interior called the plains division and hilly area of the Eastern Ghats flanking it on the North and West called the Agency Division. The Agency Division consists of the hilly regions covered by the Eastern Ghats with an altitutde of about 900 metres dotted by several peaks exceeding 1200 metres. Sankaram Forest block topping with 1615 metres embraces the Mandals of Paderu, G. Madugula, Pedabayalu, Munchingput, Hukumpeta, Dumbriguda, Araku Valley, Ananthagiri, Chinthapalli, G.K. Veedhi, and Koyyuru erstwhile Paderu, Araku Valley and Chinthapalli taluks in entirety. Machkhand river which on reflow becomes Sileru, drains and waters the area in its flow and reflow and is tapped for Power Generation. The other division is the plains division with altitude no where exceeding 75 metres watered and drained by Sarada, Varaha and Thandava Rivers and revulets Meghadrigedda and Gambheeramgedda.

Since no major Irrigation system exists significant sub regional agronomic variations exist in this division. Along the shore lies a series of salt and sandy swamps. The coast line is broken by a number of patches of barren lands, the important of them being the Dolphin's Nose which had afforded the establishment of Natural Harbour at Visakhapatnam, Rushikonda(v) Polavaram Rock and the big Narasimha Hill at Bheemunipatnam. Administratively, the District is devided into three (3) Revenue Divisions and forty three (43) Mandals.

iii. Demographic Characterists

The population of the district is 38.32 lakhs as per 2001 Census and this constituted 5.0% of the population of the state while the Geographical area of the District is 11161 Sq. KM. which is only 4.1% of the area of the State. Out of the total population 19.30 lakhs are Males and 19.02 lakhs are Females. The Sex Ratio is 985 Females per 1000 Males. The District has Density of population of 343 per Sq.Km. Agency area shows lesser Density and plain area higher density. 39.90% of the population resides in the ten (10) Hierarchic urban settlements while rest of the population is distributed in 3082 villages. Scheduled Castes constituted 7.82% of the population while Scheduled Tribes account for 14.55% of the population of the district. The district has a work force of 16.03 lakhs constituting about 41.83 of the population besides the marginal workers to a tune of 2.97 lakhs as per 2001 Census. The cultivators constitute 36.31% Agricultural Labourers 23.60% and the balance of 40.09% engage in Primary, Secondary and Territory sectors as per 1991 census.

iv. Education

There are 20.02 lakhs literates forming 52.25% of the total population of the District. Male literates constitute 30.56% while female literates forming 21.69%.

v. Climate

The district has diverse climatic conditions in different parts of it. Near Coast the air is moist and relaxing, but gets warmer towards the interior and cools down in the hilly areas on account of elevation and vegetation. April to June is warmest months. The Temperature (at Visakhapatnam Airport) gets down with the onset of South West Monsoon and tumbles to a mean minimum of 18.8º C by December after which there is reversal trend till the temperature reaches mean maximum of 37.4º C by the end of May during 2002-2003.

The District receives annual normal rainfall of 1202 mm.of which south-west monsoon accounts for 53.9% of the normal while North-East monsoon contributes 24.8% of the normal rainfall during 2001-2002. The rest is shared by summer showers and winter rains. Agency and inland Mandals receive larger rainfall from the Sourth West Monsoon, while Coastal Mandals get similarly larger rainfall from North-East monsoon. But both the monsoons play truant, variations of South-west monsoon accounting for 15.3% of normal and North-west monsoon to 33.2% of normal. Since the variation for most periods is on the negative side of log `Y' and since even the years of normal rainfall are characterised by long dry spells during one or more parts of the crop season, the district experiences drought conditions too often, as no major irrigation system exists to cushion the vagaries of the monsoon.

Research Methodology 127

vi. Soils

Red Loamy soils predominate with coverage of 69.9% of the villages of the district. The Soils are poor textured and easily drained. Sandy loamy soils come next with 19.2% villages coverage, largely confined to the coastal areas of Nakkapalli, Payakaraopeta, S.Rayavaram, Rambilli, Atchutapuram, Paravada, Visakhapatnam, Pedagantyada, Gajuwaka and Bheemunipatnam Mandals and to certen streches in the interior Mandals of Chodavaram, Narsipatnam, K.Kotapadu and Madugula. Black cotton soils come up next having sizeable chunks of area in K.Kotapadu, Devarapalli, Cheedikada, Paderu and Hukumpeta Mandals. 45% of the soils in the district are low in organic content and 55% in Phosphorous content.

vii. Land Use

The total geographical area of the district is 11.34 lakh hectares of this 30.5% alone is arable area while 42.1% is forest area. The rest is distributed among "Barren and uncultivable land" about 11.6% and "Land put to non agricultural uses" about 8.9%. Out of the arable area, the net area sowed form 24.4% while cultivable waste and fallow (current and old) lands constitute about 6.4% during 2002-2003.

viii. Flora and Fauna

More than the one third of the area in the District is covered by forest. The forests are of moist and dry deciduous type. The common species available in them are Guggilam, Tangedu, Sirimanu, Kamba, Yagisa, Nallamaddi, Gandra, Vepa etc.

Bamboo shurbs are sparsely scattered. But forest area in the district has been showing a quiescent decline since 1955-56 perhaps due to podu practice, indiscriminate grazing and browsing. To stem this, regeneration programmes are being carried out. Chinthapalli Teak Plantation is an off shoot of this. The latest caper in this regeneration programme is rising of Teak, Silver trees, coffee plantations, as the agency areas are found suitable agronomically for coffee growth. Coffee plantations have been raised in about 10000 Acres in Chinthapalli, Minimuluru, Devarapalli and Ananthagiri regions by different agencies for different purposes. By the forest Department to conserve soil, by the Coffee board to evolve cultures suited to non-traditional areas and by the Girijan Corporation and the Integrated Tribal Development Authority (ITDA). to wean out tribals from the pernicious practices of "Podu Cultivation."

Regarding fauna the district has a livestock of 13.43 lakhs as per 1999 livestock Census. In the Livestock, Cattle form 33.4% Buffaloes 31.28% Sheep 14.3% and Goats 17.6% about wild fauna Boars and Bisons are found in Forest areas of the district and isolated instances of Cheetahs and tigers.

ix. Agriculture

Agriculture is the main stay of nearly 70% of the households. Though Visakhapantam city is industrially developing, the rural areas continued to be backward. Rice is the staple food of the people and Paddy is therefore the principal food crop of the district followed by Ragi, Bajra and Jowar and Cash Crops such as Sugarcane, Groundnut, Sesamum Niger and Chillies are important. Since there is no Major Irrigation system, only about 30% of the cropped area is irrigated under the Ayacut of the Medium Irrigation System and Mimnor Irrigation Tanks. The rest of the cultivated area is covered under dry crops depending upon the vagaries of the monsoon. The productivity of the crops is low.

x. Animal Husbandry

Animal Husbandry is an important allied economic activity to Agriculture. Next to draught animals which are main source of energy for Agriculture, Milch Animals, Sheep and Goat are important for income generation of the rural households. A sizable number of households earn subsidary income by selling milk to Visakha Dairy and in Local markets. The total livestock of the district is 13.43 lakhs of which working animals account for 2.71 lakhs while milch Animals account for 3.36 lakhs. Goats and Sheep totaling up to 4.29 lakhs are important for the livelihood of the considerable population.

xi. Fishing

It is another important economic activity of the fishermen population living in about 59 fishery villages and hamlets on coastline stretching to a length of 132 KMs. covering eleven (11) coastal mandals. About 13,000 fishermen families to eke out their livelihood from marine, Inland, and brakish water fishing besides catching fish living around Thandava and Raiwada reservoyers.

xii. Minerals

The District has mineral deposits of Bauxite Apatite (Rock Phosphate) Calcite, Crystaline limestone confined to tribal tracts. Bauxite deposits at Sapparla, Jerrila and Gudem of G.K.Veedhi Mandal are considered to be the largest in the country. Bauxite deposits are also identified at Galikonda, Katuki, Chittemgodndi of Araku group deposits, Katamrajukonda of Gurthedu sub-group of deposits. Phosphate Apatite is avilable in Kasipatnam village of Ananthagiri mandal. Rich deposits of Crystaline limestone and Calcite are mapped in Borra Caves and along the Valley up to Araku from Borra and around Valasi village of Ananthagiri mandal. Ruby Mica is another mineral available in the District essential for electrical and electronic industries. The minerals are available in the form of Phologopite and are confined to Borra tract.

Quartz is another mineral found mostly in Bheemunipatnam, Padmanabham, Devarapalli, K.Kotapadu and Ananthagiri mandals. Vermiculate found near Kasipatnam of Ananthagiri mandal. Clay deposits near Malivalasa of Araku mandal are identified. Limeshell useful for manufacture of chemical grade lime is also available in the district. Red and Yellow ochre deposits are also identified in Araku and Ananthagiri mandals.

xiii. Industries

Industrial Development is conspicuous in Visakhapatnam urban agglomeration with the large scale industries like Hindustan Shipyard, Hindustan Petroleum Corporation, Coromandal Fertilisers, Bharat Heavy Plates and Vessels, L.G.Polymers Ltd.,

Hindustan Zinc Plant and the recent giant Visakhapatnam Steel Plant and a host of other ancillary Industries. The Visakhapatnam Steel Plant is the biggest with an authorised share capital of Rs.7466 crores with a licensed annual capacity of 2.8 Million Tonnes of salable steel 3.0 Million Tonnes of Pig Iron and 8.32 lakhs Tonnes of By product. About 25,000 persons expected to be employed. The project has provided employment to 16300 persons. On the country side the agro based industries like Sugar Factories, Jute Mills and Rice Mills are there besides brick and tile units. The District has one thousand and sixty three (1063) registered factories under factories Act functioning with a working force of about seventy seven thousand two hundred three (77203) persons during 2002-2003.

xiv Communications

The District has a Road length of 7336 kms. and the National Highway 5 runs all along the coast line.

There are six hundred and sixty (660) Post Offices, seven (7) Telegraph Offices and ninety six (96) Telephone Exchanges with one lakh fifty six thousand nine hundred and ninety three (156993) telephone connections in the District..

xv. Educational and Medical Facilities

There are three thousand five hundred and fifty (3550) Primary Schools with 2.80 lakhs children on enrolment, four hundred eighty nine (489) Upper Primary Schools with an enrolment of 1.28 lakhs four hundred forty seven (447) High Schools with 2.04 lakhs pupils on roll, one hundred ninety six (196)

Junior, Degree and Professional Institutions with 0.83 lakhs students during 2002-2003.

Regarding Medical facilities, there are one hundred fifty nine (159) Government Hospitals and dispensaries both Allopathic and Indian Medicine with two thousand eight hundred nineteen (2819) bed-strength and five hundred ninety six (596) Doctors.

Section II: Profile of Andhra University

Andhra University is one of the oldest universities in the country. It was constituted way back in the year 1926. The following is the description on the profile of Andhra University, which was sourced from its website.

Andhra University was constituted by the Madras Act of 1926. The 82-year-old institution is fortunate to have Sir C.R. Reddy as its founder Vice-Chancellor, as the steps taken by this visionary proved to be fruitful in the long run.

Former President of India Dr. Sarvepalli Radhakrishnan was one of its Vice-Chancellors who succeed Dr. C. R. Reddy in 1931. The University College of Arts was inaugurated on 1st July, 1931. The inaugural courses were Telugu language and literature, History, Economics and Politics. A year later, the College of Science and Technology came into being with Honours courses in Physics and Chemistry. The University pioneered in introducing many new courses in Science, Arts, Management and Engineering in the country.

The leaders of the university always believed that excellence in higher education is the best investment for the country and engaged the services of famous educationists such as Dr. T.R. Seshadri, Dr. S. Bhagavantham, Professor Hiren Mukherjee, Professor Humayan Kabir and Dr. V.K.R.V. Rao, to mention a few who set high standards in teaching and research. Nobel Laureate Sir.C V Raman was the proud alumnus of the University and closely associated in laying research foundations in Physics. Padmavibhushan Prof. C R Rao, the renowned statistician of the world, was also the proud alumnus of the University.

Keeping in pace with the global needs and challenges under the leadership and guidance of successive Vice-chancellors, the University is offering several new Courses of relevance and Contemporary significance.

Ever since its inception in 1926 Andhra University has an impeccable record of catering to the educational needs and solving the sociological problems of the region. The University is relentless in its efforts in maintaining standards in teaching and research, ensuring proper character building and development among the students, encouraging community development programmes, nurturing leadership in young men and women and imbibing a sense of responsibility to become good citizens, while striving for excellence in all fronts.

As a sequel to this, the University has always subjected itself for continuous self-evaluation for maintaining standards and to reach set targets. Further, to assure quality in Higher Education the University has gone through the process of assessment and accreditation by National Assessment and Accreditation Council (NAAC) in April 2002. The Peer Committee has analyzed the strengths and weaknesses of the institution and has assessed the University with 'A' grade awarding it with the best percentage among the State Universities of Andhra Pradesh. The University designed and implemented Quality Management System successfully and became the first general University in the country to get ISO 9001: 2000 Certification in 2006.

The University Presently is offering three hundred and thirteen (313) Courses in Arts, Commerce, Management, Science & Technology, Engineering, Law, Pharmacy and Education. The University has five constituent colleges and four AU Campuses. The Colleges of Arts and Commerce is the biggest constituent college in the University with twenty six (26) Departments offering forty two (42) courses including four Diploma Courses. The College of Science and Technology has twenty one (21) Departments, which offer sixty three (63) Courses including one PG Diploma. The College of Engineering has fifteen (15) Departments offering Undergraduate, Postgraduate and Research Programmes. The College of Law has been identified as an advanced Center in Law by UGC. The college of Pharmaceutical Sciences is first of its kind in South India, which is offering one UG Programme and six PG Programmes, besides Research Programmes leading to Ph.D. Degree. The AU campus at Kakinada (East Godavari District) has five Departments and the AU Campus at Etcherla (Srikakulam Dist) has Eleven Departments. The AU Campus at Tadepalli Gudem has four departments and AU Campus at Vizianagaram has five Departments.

The University is also having student services and welfare centres in the campus like post and Telegraph office, Banking facilities, University employment and Guidance bureau, training and placement centre and a sports complex including a Gymnasium Hall.

Presently the University is catering to the educational needs of five green districts of Andhra Pradesh namely Visakhapatnam, East Godavari, West Godavari, Vizinagaram and Srikakulam.

The University has started school of Distance Education in the campus in the year 1972. It offers Courses ranging from Certificate Courses to Post Graduate and Professional Courses. The School of Distance Education is presently offering four (4) UG Programs, eighteen (18) PG Programs, five (5) Certification Courses, nine (9) PG Diploma Courses, (15) Professional Courses and (7) Collaborative Courses through Distance Education Mode. The School has thirty five (35) study centres spread across the State. The enrolment in UG, PG, and Professional Programmes in the School of Distance Education is about eighty thousand (80,000) students. The School has twenty three (23) permanent faculty members. It is Equipped with the latest infrastructure.

The University has well experienced and expert faculty known world wide for their research contributions. There are three hundred and fifty four (354) Professors, one hundred ninety eight (198) Associate. Professors and one hundred and fifteen (115) Asst. Professors serving the University. Organization of International and National events is a regular feature of the University. There is more than two thousand five hundred (2500) support staffs working in the University.

The alumni of the University occupy important positions in government administration, Industry and research organizations with in and out side the country. The Government of Andhra Pradesh has appointed the professors of the University as chairman and vice-chairman of the APSCHE and presently ten Universities had Vice-chancellors from Andhra University.

In the next chapter the data was analysed and the necessary consequential inductive and deductive logical premises are arrived at and accordingly the results were reported.

> "I don't want my house to be walled in on all sides and windows to be stuffed. I want the culture of all lands to be blown about my house ass freely as possibly but I refuse to be blown off my feet by any."
>
> _____ Mahatma Gandhi

Chapter – 5

Abstract

In this chapter, data collected was analysed utilizing the tools of research; inferences were drawn and explained in detail. The aspects of socio-economic background of students, parents, environment and ethical problems and their responses from the students and its ramifications were clearly analysed with the help of inductive and deductive logical principles.

ANALYSIS OF DATA AND PRESENTATION OF REPORT

Analysis of Data and Presentation of Report   137

"Education means an all round drawing out of the best in child and in men, in body and mind, and in spirit, thus lifting him to live a suitable life fit to face challenges"

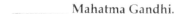 Mahatma Gandhi.

In this section an attempt was made to understand the opinions expressed by the students, on issues pertaining to ethical practices and the prevailing environment in the educational institutions. The responses received from the sample selected for the students were analysed and the outcome is discussed in this chapter. Since the sample consists of students, in order to make better understanding on the issues concerned, the analysis of data pertaining to the category of sample is presented in this section.

Students – Response:

As described in the research methodology adopted for the study, a sample of 300 students was selected who were pursuing higher education in Visakhapatnam city. Further, in order to understand the impact of stream of education i.e. specific branches of education pursued, a sample of 100 students from each of the distinct streams of education were selected. The distribution of sample is presented in the following Table 1.

Table: 5.1: Distribution of Sample

Sl. No.	Category of Higher Education	Number of students selected
1	Science & Technology	100
2	Commerce and Management	100
3	Arts and Social Sciences	100
	Total Sample	300

Thus, a sample of 300 students was selected for the study who are pursuing higher education i.e. graduation and post graduation, in Visakhapatnam city of Andhra Pradesh in India. Under the category of Science and Technology, students pursuing graduate and post graduate courses in science and technology (B.Sc., B.E/B.Tech etc) were selected. Under the stream of Commerce and Management (B.Com, M.Com, M.B.A etc) were selected. Under the stream of Arts and Sciences (BA, MA etc) were selected for the purpose of the study.

Data from the students were collected by using a Questionnaire having different details on socio-economic particulars of students, their perception and opinion on ethical practices and environment in educational institutions. The responses received in this regard were analysed and presented in the following sections.

a. Socio-economic particulars

In this sub-section, particulars pertaining to socio-economic conditions prevailing among the students are presented.

i. Community and Gender Profile of Students

Analysis of Data and Presentation of Report

Table: 5. 2: Community and Gender Profile of Students

S.No.	Category of Students	Sex		Total				
		Male	Female	SC	ST	BC	OC	Total
1	Science and Technology	68	32	18	6	32	44	100
2	Commerce and Management	61	39	21	9	36	34	100
3	Arts and Social Sciences	64	36	20	8	37	35	100
	Total	193 (64.33)	107 (35.67)	59 (19.67)	23 (7.67)	105 (35)	113 (37.66)	300

Note: Figures in parenthesis indicate percentages

The analysed data in this regard is presented in Table 2. As observed from the data presented, majority of them 193 students (64.33%) were male and female represent the remaining. In fact, given the general prevailing ratio of students in higher education, the distribution of male and female is more or less in tune with the existing conditions. Hence, it may be concluded that the randomly drawn sample of 300 students represent the actual ratio prevailing in the higher education institutions. The distribution of sample across different streams of education is also more or less equal.

In regard to community profile, almost equal numbers of students were from backward community 105 students (35%) as well as Other Castes or Forward Caste students 113 (37.66%) community. Among the remaining students, communities of Scheduled Castes 59 (19.67%) and Scheduled Tribes 23 (7.67%) were quite dominant. Thus, the distribution of sample in terms of community profile is also considered to be in tune with the prevailing conditions. As backward castes and forward caste communities dominate the society in terms of numbers, the sample also carried relevant representation. Further, the students from Scheduled castes and Schedule Tribes were also in tune with their actual distribution in the society. The same is the case with other communities. Thus, the randomly drawn sample of students represents the true picture of social communities prevailing in the society.

ii. Occupational Profile

Table: 5. 3: Occupational Profile of Students' Families

Sl. No.	Category of Students	Occupational Profile						Total
		Agriculture	Govt. Ser.	Business	Artisans	Agrl. Labr.	Non Ag. Lab.	
1	Science and Technology	21	31	37	6	3	2	100
2	Commerce and Management	9	39	42	4	4	2	100
3	Arts and Social Sciences	35	27	25	5	4	4	100
	Total	65 (21.67)	97 (32.33)	104 (34.67)	15 (5)	11 (3.67)	8 (2.66)	300

Note: Figures in parenthesis indicate percentages

While understanding the socio-economic profile of the students, it is essential to understand the occupational profile of the students' family. Hence, relevant data in this regard is provided in Table 3. As may be observed from the data presented, the occupational background of the students' families is dominated by Business (34.67%) and Government Service (32.33%). In fact, two thirds of the student's parents (67%) are having the occupational background of either business or government service. In fact, since the sample for the study was drawn from the students who were

Analysis of Data and Presentation of Report   141

pursuing higher education, it is quite natural to expect the domination from these two occupations given the expenditure involved. The next major occupation prevailing among the students is agriculture (21.67%) and then followed in a nominal way in reference to artisans (5%), agriculture labour (3.67%) and non agriculture labour (2.66%). Since the Visakhapatnam city is the focal point for higher education, it meets the needs of rural and tribal areas people inhabiting in the hinterland. Since, especially in rural and tribal areas, the major occupation is agriculture and thus it registered substantial presence among the students.

With reference to different streams of education, an interesting trend may be observed. In Arts and Social Sciences, the major occupational background is dominated by agriculture when compared to other streams of education. As the streams of Science and Technology as well as Commerce and Management attracts more students of urban background and also attracts more expenditure, the rural based students would certainly lag behind in other occupations in this regard. Similarly, students from the occupational background of government service and business were present more under the streams of science and technology as well as commerce and management when compared to arts and social sciences. In this context too, it may be pointed out that, those parents who are in government service and business are in a better position to send their children for higher education in science and technology as well as commerce and management.

Thus, the occupational profile of students' families and its distribution across different streams of education represent the conditions prevailing in the society.

iii. **Profile of Annual Income**

While understanding the socio-economic conditions prevailing among the students, it is also essential to pay attention on the financial background of students' families. Hence, data pertaining to annual income of students' families is presented in Table 4.

Table: 5.4

Profile of Annual Income of Students' Families

S. No.	Category of Students	Annual Income				Total
		Up to Rs. 50,000	Rs. 50,000 – Rs. 1,00,000	Rs. 1,00,000 – Rs. 3,00,000	Rs. 3,00,000 – Rs. 5,00,000	Above Rs. 5,00,000	
1	Science & Technology	3	28	45	16	5	100
2	Commerce & Management	12	36	42	7	3	100
3	Arts & Social Sciences	21	48	20	6	5	100
	Total	39 (13)	112 (37.33)	107 (35.67)	29 (9.67)	13 (4.33)	300

Note: Figures in parenthesis indicate percentages

Among the 300 students selected for the study, slightly higher than one third of students (37.33%) were having an annual income up to Rs 1 lakh per annum and then closely followed by 107 students (35.67%) who reported having an annual income up to Rs 3 lakhs. Almost one tenth of the students (9.67%) reported that their annual income is up to Rs 5 lakhs and only 13 students (4.33%) reported that they were having annual income higher than Rs 5 lakhs. The remaining 39 students were having an annual income up to Rs 50,000/- who mostly belongs to poverty group.

Since higher education involves more expenditure, it is expected that poor people can not afford the same and thus the representation of students with poverty background was quite less. Those who were pursing the higher education in this regard must be accessing certain fee concessions and benefits provided to them by the government. Thus, the overall distribution of sample in reference to annual income is quite in tune with the economic conditions prevailing in the society.

Further in reference to different streams of education, it may be observed that students having an annual income up to Rs 1 lakh were mostly found in arts and social sciences stream since it attracts lesser expenditure. Similarly, higher income families naturally prefer education for their children in streams like science and technology as well as commerce and management. The distribution of sample in this regard is also in conformity of their choice. To sum up, nearly 50% of the sample were having an annual income up to Rs 1 lakh and most of them were distributed in arts and social sciences stream. However, it may be pointed out that government is providing assistance as well as certain benefits to poor people for their children education and this in background, though substantial expenditure is involved in higher education, poor families could able to send their children to higher education. Similarly, the higher income group prefer engineering and technology as well as commerce and management streams of education for their children. All these observations are quite in tune with the grass-root realities.

iv. **Students' Place of origin**

In terms of ethical practices and while going through the educational environment, the place of students' origin also matters most. Hence, an attempt was made to understand the place of origin of students. The details in this regard are presented in Table 5.

Table: 5.5
Students' Place of Origin

S.No.	Category of Students	Students' Place of Origin				Total
		Urban	Rural	Semi-urban	Tribal	
1	Science & Technology	41	20	30	9	100
2	Commerce & Management	32	29	35	4	100
3	Arts & Social Sciences	28	36	31	5	100
	Total	101 (33.67)	85 (28.33)	96 (32)	18 (6)	300

Note: Figures in parenthesis indicate percentages

It may be observed from the data presented that most of the students were having either urban (33.67%) or semi-urban (32%) background. The remaining students were from rural areas (28.33%) in a substantial manner and very few of them (6%) from tribal areas. Given the reservation policy pursued, the presence of students from tribal areas is quite expected in Visakhapatnam city. Further, on close observation of students pursuing different streams of higher education, it may be seen that substantial number of students from the rural areas were found pursing arts and social sciences courses rather than the other streams of higher education. On the other hand, the urban and semi-urban students were found more in the streams of engineering and technology as well as commerce and management. Unlike the primary and high school levels of education, wherein the urban and rural differentiation is not much, the higher education attracts more students with urban or semi-urban background by virtue of their exposure and vicinity to the higher education facilities. Hence, most of the sample of students selected for the study was having their origin from urban and semi-urban areas and this has been in tune with prevailing socio-economic conditions in the society.

iv. Distinction held by students in eligible examination

In terms of ethical practices and understanding the environment prevailing in educational institutions, it is essential to understand the performance of students in their eligible courses prior to pursuing higher education. Hence, data pertaining to level of distinction held by the students in their eligible examinations concerned is obtained and provided in Table 6.

Table: 5.6

Marks Scored By Students in their Qualifying Examinations

S.No.	Category of Students	Percentage of Marks Obtained in Qualifying Examinations					Total
		90% and above	80% to 90%	70% to 80%	60% to 70%	Less than 60%	
1	Science & Technology	29	36	20	10	5	100
2	Commerce & Management	22	21	41	9	7	100
3	Arts & Social Sciences	9	12	51	18	10	100
	Total	60 (20)	69 (23)	112 (37.33)	37 (12.33)	22 (7.34)	300

Note: Figures in parenthesis indicate percentages

It may be observed from the data presented that one fifth of the students (20%) were reported scoring more than 90% in their eligible examination concerned. However, the distribution of sample in this regard is not equal when different streams of higher education were considered. Among these students, most of them were from science and technology stream (29 students) and then followed by commerce and management (22 students). Only 9 students were found under the stream of arts and science. Thus, those held highest bracket of marks were opting for higher education streams in the fields of science, engineering, commerce and management.

Similar is the distribution of sample with reference to students who secured marks between 80% - 90% (23%). At the overall sample level, majority of the students obtained marks in the range of 70%-80% (37.33%) and in this regard the distribution of students was more in the streams of arts and social sciences (51 students) and commerce and management (41 students) when compared to science and engineering stream (20 students).

There were 37 students (12.33%) who reported scoring 60% - 70% and the remaining 22 students (7.34%) were reported having less than 60% marks in their qualifying courses. In this regard too, it may be observed that most of these students were distributed in arts and social sciences stream. To sum up, it may be pointed out that the students who secured higher marks in their qualifying examinations were mostly found in science and engineering stream and then followed by commerce and management. Students who

Analysis of Data and Presentation of Report   147

secured comparatively lesser marks were found in arts and social sciences stream.

To conclude with on socio-economic conditions of the students selected for the study, it may be observed that the sex-ratio among the sample of students has been in tune with the representation of boys and girls in higher education system. Given the reservation of 33% for girl students in higher education, the distribution of sample has also been more or less in similar ratio. Similar is the case with community profile of students which has also been in tune with prevailing community profile in the society. Thus, the randomly selected sample 300 students represent the true picture of higher education system.

The occupational background of the students' families is dominated by government service as well as business and then followed by agriculture. Given the profile of higher education, the occupational background of the students has also been in conformity with the same. Similar is the case with annual income of students' families. The middle income-group of annual income (Rs 1 lakh to Rs 3 lakhs) is more dominant as well as lower income level between Rs 50,000 – Rs 1 lakh. Since the society is dominated by lower and middle income groups, the students' family profile also reflects the same. Further, those with urban and semi-urban background dominated the higher education scenario in the study area and this has been the usual case everywhere. Similarly, those who have scored higher marks in the qualifying examinations concerned were found distributed more in science and engineering stream and then followed by commerce and management and arts and social sciences stream of students. This has also been in tune with the usual trend observed in the higher education scenario.

Thus, the sample of the study is in close proximity with the conditions and trends prevailing in the society and thus represents the general scenario.

Table: 5.7: Students Awareness on Ethical Practices

Sl. No.	Statement	Yes	No	No idea
1.	Have you heard about ethical practices in higher education	133 (44.3)	130 (43.3)	37 (12.3)

The investigator enquired about the respondents about their awareness on ethical practices. Out of 300 respondents only 133 (44.3%) have agreed that they have knowledge on ethical practices. But 167 (45.6%) respondents have agreed that they are ignorant about this. This informs even the students studying at degree level have no knowledge about ethical practices. In this regard the planners of education have to endeavour and take steps to ensure that all the students are having reasonable level of ethical attitude is built up ensuring the total awareness of ethical concepts.

Opinion on importance of ethical practices in higher education.

Irrespective of your knowledge on ethical practices in higher education, it is essential to keep seat of higher education with certain principles. Do you agree?

Analysis of Data and Presentation of Report   149

Table: 5. 8

Student's Opinion on Ethics in Higher Education.

Sl. No.	Agree	Number	Percentage
1.	Yes	227	75.7
2.	No	73	24.3
	Total	300	100.0

The response from the statement that it is essential to keep the seat of higher education with certain principals majority 227(75.7%) agreed. The result indicates though the students are inclined towards ethical practices and principles the management of higher education can keep an eye in this regard and also ensure the balance of 73 (23.3%) students will follow the majority. As it is a fact that even highly ethical people succumb to the demands of needs and sacrifice ideals to meet the immediate needs in the society.

Table: 5.9

Respondents views on Ethical Practices Ethical Practices.

Sl. No.	Statement	Strongly Agree	Agree	Neutral	Slightly Disagree	Strongly Disagree
1.	Does your institution promises to have such atmosphere	18 (6.0)	23 (7.7)	47 (15.7)	117 (39.0)	95 (31.7)
2.	It is not the achievement in examinations but the way process of education is being organised in your institution matters most. Do you agree?	95 (31.7)	50 (16.7)	30 (10.0)	24 (8.0)	101 (33.7)
3.	Are you satisfied with the practice of moral educational practices in your institution?	50 (16.7)	36 (12.0)	60 (20.0)	56 (18.7)	98 (32.7)
4.	Do you appreciate the opinion that ethical practices in educational institutions are an asset for building a strong value system?	78 (26.0)	32 (10.7)	82 (27.3)	88 (29.3)	20 (6.7)

Analysis of Data and Presentation of Report 150

Sl. No.	Statement	Strongly Agree	Agree	Neutral	Slightly Disagree	Strongly Disagree
5.	In your opinion what is your satisfaction level with reference to ethical practices in your institution?	127 (42.3)	56 (18.7)	31 (10.3)	12 (4.0)	74 (24.7)
6.	Do you think that the students themselves have a role to play in strengthening ethical practices in educational institutions?	132 (44.0)	61 (20.3)	32 (2.3)	7 (2.3)	68 (22.7)
7.	How do you rate the teaching faculty of your institution in terms of their adherence to ethical practices?	82 (27.3)	27 (9.0)	15 (5.0)	11 (3.7)	165 (55.0)
8.	Do you think that the management of your institution is paying proper attention in maintaining ethical practices?	114 (38.0)	32 (10.7)	69 (23.0)	28 (9.3)	57 (19.0)
9.	Do you think that the ethical practices, in its overall terms, is being maintained quite well?	142 (47.3)	76 (25.3)	38 (12.7)	12 (4.0)	32 (10.7)
10.	Irrespective of the status of ethical practices in your institution, it is quite important to understand that all stakeholders do have equal responsibility in maintaining ethical practices?	119 (39.7)	36 (12.0)	20 (6.7)	4 (1.3)	121 (40.3)
11.	Given your opinion and experience in the institution in reference to ethical practices, how do you consider its role in building your future career?	94 (31.3)	42 (14.0)	8 (2.7)	23 (7.7)	133 (44.3)
12.	If a person has the intention and intensiveness to adhere to ethical practices, no one can stop him from practicing it.	140 (46.7)	27 (9.0)	59 (19.7)	7 (2.3)	67 (22.3)
13.	Intention to practice ethical practices is an innate behaviour rather than an induced behaviour.	115 (38.3)	77 (25.7)	24 (8.0)	56 (18.7)	28 (9.3)
14.	Adherence to ethical practices is a trait which emanates from family background rather than the educational institution.	107 (35.7)	66 (22.0)	47 (15.7)	42 (14.0)	38 (12.7)
15.	Irrespective of level of education, ethical practices have to be practiced to build strong educational career.	106 (35.3)	51 (17.0)	27 (9.0)	20 (6.7)	96 (32.)
16.	Educational institutions build-up their reputation based on ethical practices rather than the educational achievements.	132 (44.0)	58 (19.3)	43 (14.3)	28 (9.3)	39 (13.0)
17.	In reference to your own educational achievement, your adherence to ethical practices played a vital role.	141 (47.0)	63 (21.0)	20 (6.7)	4 (1.3)	72 (24.0)
18.	Your social and economic background has something to do with your intention to adhere to ethical practices.	117 (39.0)	46 (15.3)	39 (13.0)	12 (4.0)	86 (28.7)

Analysis of Data and Presentation of Report   151

The analysis of the above table with reference to the questions and their responses is presented as following:

1. The respondents were asked regarding ethical practices in the form of eighteen statements and their opinions were recorded. Regarding the statement 'does your institution promises to have such atmosphere' only 41 (13.7%) students positively admitted but majority 212 (70.7%) students have expressed negative opinion with the statement. It was found from the analysis that the ethical practices are not being given their importance by the managements of the institutions.

2. The respondents have more positively inclined towards process of education in the institutions and 145(48.4%) preferred this method.

3. Satisfaction towards the practice of moral and educational practices in their institutions 154 (51.4%) respondents admitted that they are dissatisfied with the existing practices. This shows the apathy of the managements towards implementation of value based education.

4. Majority of the respondents 120(36.7%) feels that ethical practices in educational institutions are an asset for building a strong value system but nearly equal number of respondents 108(36%)have negative opinion. The managements have to build a strong environment to influence the minds of the students.

5. With regard to respondents satisfaction level about the practicing ethical values in the institutions majority 182 (51%)

have expressed their satisfaction. This shows in many institutions the existing implementation of ethical practices is worth to note.

6. Majority of the students 193(60.3%) have expressed positive ness that the role of the students is important. This indicates the respondents had self determination and self reliance.

7. In view of the statement faculty's adherence to ethical practices majority 176(58.7%) students have negative opinion and only 109 (36.3%) students either strongly agreed or agreed in this regard. There is an urgent need on the part of the management of the institutions to look into the matter and inculcate the practice of ethical values on their job situation.

8. Majority 146 (48.7%) of the respondents have positive ness towards attention of the management in creating positive environment towards ethical practices.

9. In regarding maintenance of ethical practices in their study environment 142(47.3%) students strongly agreed and 76(25.3%) students agreed and only 34(14.7%) students have negated with the statement.

10. A marginal majority of respondents' 155(51.7%) feels that ethical practices in the institutions are the equal responsibility of management and students. But 121(40.3%) have strongly disagreed about equal responsibility. There is a need to create awareness with the students that joining hands for the noble cause is the responsibility of both the students and the heads of the institutions.

11. The responses indicate that 94 (31.3%) students strongly agreed to this aspect helping them for their future career but majority 133(44.3%) students disagreed with this statement. This indicates the respondents are not fully dependent on ethical practices and they have other plans also in planning their future career.

12. Majority of the students 140(46.7%) strongly agreed that the influence of others have no impact if the person have self determination. Simultaneously 67(22.3%) respondents have an opinion that influence plays a role in the institutions, proper capacity building training programmes are to be organized to strengthen the attitude of the pupil.

13. Majority 192(64%) students are either strongly agreed or agreed that the ethical practices will be in built in character rather than forced one. From the opinion of the students the system of ethical practices is inherent quality of human beings.

14. Similarly regarding statement of ethical practices is a trait which comes from the family background, majority 167(57.7%) expressed their positive ness rather than other respondents 80(26.7%) have negative assumption. Many students have an opinion that ethical practices will be inculcated and affect from the family environment rather than institutional environment.

15. As regards the social background affecting the adherence to ethical practices 163 (54.3%) expressed their agreement with

different degrees in the positive and affirmed it. Thirty nine respondents 39(13%) have expressed that they have no idea and 98(32.7%) affirmed in the negative that the economic background does not affect their adherence to the ethical practices. **

16. Majority 190(63.3%) students agreed that the credibility and reputation of the educational institutions **enhanced** only on the ethical practices adopted by the administration rather than students academic performance.

17. Regarding student's opinion that their academic performance is to a depended upon the way their adherence on ethical practices, in this aspect majority 204(68%) students either strongly agreed or agreed with the statement in the affirmative. Only 72(24%) disagreed that personal practice if ethical values have no relation to his academic performance.

18. Majority 163(54.3%) expressed positive opinion that their social and economic background had an impact on their grooming attitude to ethical practices, only 12(4%) negated with this statement.

Analysis of Data and Presentation of Report

Table: 5.10:

Students – Environment - Ethics and Higher Educational Institutions.

Sl. No.	Statement					
1.	The social and economic milieu you are exposed in the educational institution is not suitable for ethical practices.	112 (37.3)	76 (25.3)	26 (8.7)	32 (10.7)	54 (18.0)
2.	Though you have strong personal intention to adhere to ethical practices yet the 'commercial environment' in educational institutions is dampening your spirits?	98 (32.7)	62 (20.7)	59 (19.7)	65 (21.7)	16 (5.3)
3.	In the light of commercialisation of educational institutions, the principle of ethical practices is sacrificed.	134 (44.4)	31 (10.3)	22 (7.3)	34 (11.3)	79 (26.3)
4.	The physical environment in your institution is not conducive to ethical practices.	123 (41.0)	31 (10.3)	26 (8.7)	16 (5.3)	104 (34.7)
5.	The academic atmosphere in your institution is not conducive to ethical practices.	107 (35.7)	36 (12.0)	67 (22.3)	83 (27.7)	7 (2.3)
6.	The inter-personal relations among the students matters most in ensuring adherence to ethical practices.	124 (41.3)	54 (10.8)	61 (20.3)	16 (5.3)	45 (15.0)
7.	The student-academic staff relations are quite essential to ensure ethical practices in educational institutions.	125 (41.7)	53 (17.7)	19 (6.3)	44 (14.7)	59 (19.7)
8.	The physical surroundings (urban/semi-urban/rural) plays role in ensuring ethical practices.	137 (45.7)	58 (19.3)	59 (19.7)	22 (7.3)	24 (8.0)
9.	The practices of 'educational counselling' is quite essential to induce ethical practices among students.	74 (24.7)	75 (25.0)	72 (24.0)	30 (10.0)	49 (16.3)
10.	The managerial practices adopted in your institutions has something to do with inducing ethical practices among students.	98 (32.7)	35 (11.7)	53 (17.7)	100 (33.3)	14 (4.7)
11.	The educational background of parents do matter in ensuring ethical practices among students.	119 (39.7)	83 (27.7)	32 (10.7)	24 (8.0)	42 (14.0)
12.	The place of self-origin like urban/semi-urban/tribal has its impact on inducing ethical practices among students.	90 (30.0)	47 (15.7)	20 (6.7)	32 (10.7)	111 (37.0)
13.	The level of education among family members has its impact on inducing ethical practices among students.	109 (36.3)	40 (13.3)	31 (10.3)	27 (9.0)	93 (31.0)
14.	It is not the management but the teaching staff of the educational institution has more influence on students while inducing ethical practices.	90 (30.0)	49 (16.3)	27 (9.0)	44 (14.7)	90 (30.0)

Analysis of Data and Presentation of Report 156

15.	Though teaching staff has influence on students yet its is the good practices from management of the educational institution that provides right atmosphere to practices ethical practices.	84 (28.0)	69 (23.0)	84 (28.0)	31 (10.3)	32 (10.7)
16.	In the realm of higher education, it is the strong moral practices bequeathed from previous level of education matters most in inducing ethical practices.	80 (26.7)	34 (11.3)	75 (25.0)	108 (36.0)	3 (1.0)
17.	Whatever the influence of various factors on ethical practices, it is the personal attention and discipline that matters most in inducing ethical practices.	150 (50.0)	70 (23.3)	30 (10.0)	30 (10.0)	20 (6.7)
18.	Irrespective of the quality of educational institution, it is the stream/branch of education matters most in inducing ethical practices.	134 (44.7)	64 (21.3)	64 (21.3)	23 (7.7)	15 (5.0)

The analysis was carried out to estimate the impact of institutional environment on the students' adherence to ethical practices.

1. Majority 188(62.6%) of the respondents either strongly agreed or agreed that different social and economic backgrounds of respondents are not congenial to follow ethical practices as they keep on preaching different lifestyles and different beliefs and attitudes. 26(8.7%) respondents expressed that they neither agree nor disagree and 86(28.7%) are negative to the statement.

2. Regarding the commercial environment in the institutions always distract and divert the strong intention to adhere to ethical practices majority of the respondents 160(50.4%) agreed with the statement.

3. With the statement that the ethical practices are sacrificed because of the commercial attitude of educational institutions 165(54.7%) either strongly agreed or agreed and only 79 (26.3%) respondents strongly disagreed with this opinion.

4. Regarding institutional physical environment is not conducive to the ethical practices 154 (51.3%) have positive opinion, but 113 (27.6%) disagree with their co students opinion.

5. Similarly majority 143 (47.7%) respondents feel that academic atmosphere in their institution is not conducive to ethical practices and about 90(30%) expressed that the atmosphere helps them to adhere to ethical practices.

6. Majority 178 (52.1%) respondents expressed their positiveness that the cordial relations among students made a way to ensuring adherence to ethical practices in institutional environment, only 51 (20.3%) respondents have negativeness in this regard.

7. The impact on students to adhere to ethical practices only 125 (41.7%) strongly agreed that positive relation with faculty yield better results to follow ethical practices during their period of study in the institution.

8. The physical surroundings play a vital role on ensuring ethical practices majority 195(65%) have expressed positive opinion.

9. The respondents 149 (49.7%) have an opinion that for effective ethical practices in educational institutions there must be a provision to establish educational counselling units.

10. The efforts of management to inculcate ethical practices in the institution environment 98(32.7%) respondents strongly agreed and were positive on the efforts of their respective institutions.

11. Two thirds 202(67.4%) expressed positive opinion that the educational background of parent have a greater influence on creating ethical practices among students.

12. The influence of birth plan on inducing ethical practices, a marginal number of respondents 111(37%) disagreed with the statement. Accordingly the results a different opinion were established.

13. Nearly half of the respondents 149 (49.6%) agreed that in inducing ethical practices the family members education has a bearing.

14. The respondents 90 (30%) disagreed with the statement, teaching staff is only responsible to induce ethical practices in the institutions.

15. There is a mixed response from the statement that in spite of staff influence on students to create a right atmosphere on the part of the management is necessary, 153(51%) respondents have strong acceptance and the others have different views.

16. In higher education while inculcating ethical practices, the practices adopted in previous level of education have an impact, a marginal majority 108(36%) disagreed with the statement.

17. Majority 220(73.3%) agreed that the personal determination and discipline only matter in practicing ethical principles despite of influence of various factors.

18. Majority 198(66%) respondents have an opinion that the stream/branch of education is more important than quality of educational institutions background in inducing ethical practices.

Table: 5.11

Socio-Economic Background - Ethical Views of Students.
Mean impact on ethical practices pose, SDs obtained by the respondents belonging to different socio economic groups and the respective t/F values.

Sl. No.	Character	Variable	N	Mean	SD	T/F Value
1.	Gender	Male	193	48.7306	15.9426	1.132NS
		Female	107	50.9252	16.3400	
2.	Caste	Forward Castes	113	51.1858	17.2686	2.683*
		Backward Castes	105	47.0952	15.7685	
		Scheduled Castes	59	48.0508	15.7255	
		Scheduled Tribes	23	56.0870	8.9996	
3.	Occupation	Agriculture	65	50.6615	14.6557	3.174**
		Govt. Service	97	46.9278	15.0272	
		Business	104	48.1538	18.3392	
		Artisans	15	58.5333	10.5076	
		Agricultural Labour	11	57.4545	8.9372	
		Non Agricultural Labour	8	61.3750	12.2817	

		Rs. 50,000/- below	39	54.3846	14.4106	
4.	Income	Up to Rs. 1,00,000/-	112	45.6786	16.9367	4.688**
		Up to Rs. 2,00,000/-	107	52.9346	14.5513	
		Up to Rs. 3,00,000/-	29	44.3793	19.1897	
		Above Rs. 3,00,000/-	13	51.2308	5.1503	
5.	Nativity	Urban	101	46.3168	15.8505	4.223**
		Rural	85	51.4118	15.3555	
		Semi-urban	96	49.3021	16.5664	
		Tribal	18	59.6111	13.8315	
6.	Stream of Higher Education	Science / Engineering	108	51.0741	15.5411	0.769NS
		Commerce Management	97	48.5464	16.4785	
		Arts / Social Science	95	48.7263	16.3437	

** Significant at 0.01 level, * Significant at 0.05 level, NS Not Significant

The information on parental occupation on students' attitudes towards ethical practices, the students from artisan families obtained on better mean score on 51.2174 than non-agriculture labor 50.6250, agriculture labor 49.4545, agriculture 46.6000, business 45.1731 and Govt-service 44.1753 obtained 'f' value is insignificant. The SD value for male and female are 15.9426 and 16.3400 respectively. It was observed from the analysis students from families belonging to fixed monthly regular income are lagging than business belongings to un-organized sector.

The impact of family annual income on students shows students from business in score at the Rs.30, 000/- per annum have more mean score of 50.1538, fallowed by incomes up to Rs.2, 00,000 between on 50,000/-(48.6154), up to Rs.1, 00, 000 (42.2946) and up to Rs 3, 00,000.

Andhra University, Visakhapatnam

Analysis of Data and Presentation of Report   161

The 'f' value is 4.365 and significant at 0.01 levels. Surprisingly the students from business income groups have more positive opinion on ethical practices.

The nativity plays a vital role to adjust to the environmental imbalances. From the analysis it was found that the tribal nativity students possess a higher mean score of 52.4444 followed by rural 48.1529, semi-urban 44.6979 and urban 43.6040 obtained values is 2.745, which is significant at 0.05 levels. The results of the table inform the more modern exposure on the part of the students will lead to change of attitude towards participating ethical values.

The attitude of the students on ethical practices was calculated depending upon their discipline/faculty. It was observed that students from science and engineering discipline acquired higher mean score of on 47.8611 than arts social science 44.6804 and 'f' value is insignificant. This can be / may be due to teaching learning environment of the class room. In nutshell, female students belong to artisans' families, higher income tribal relatively and studying science/engineering students having higher mean score than their counterparts.

Analysis of Data and Presentation of Report

Table: 5.12: Institutional Environment – Ethical Views & Socio Economic Background.

Mean impact on Environmental conditions pose, SDs obtained by the respondents belonging to different socio economic groups and the respective t/F values.

Sl. No.	Character	Variable	Number	Mean	SD	t/F Value
1.	Gender	Male	193	45.4093	15.2893	0.557NS
		Female	107	46.4299	15.0697	
2.	Caste	Forward Castes	113	46.5310	15.8391	1.518NS
		Backward Castes	105	44.8000	15.0783	
		Scheduled Castes	59	43.9322	15.9891	
		Scheduled Tribes	23	51.2174	8.0054	
3.	Occupation	Agriculture	65	46.6000	13.2037	0.989NS
		Govt. Service	97	44.1753	15.5851	
		Business	104	45.1731	16.8762	
		Artisans	15	51.4000	11.3314	
		Agricultural Labour	11	49.4545	9.5327	
		Non Agricultural Labour	8	50.6250	13.9789	
4.	Income	Rs. 50,000/- below	39	48.6154	12.5292	4.365**
		Up to Rs. 1,00,000/-	112	42.2946	16.8964	
		Up to Rs. 2,00,000/-	107	48.8692	12.7953	
		Up to Rs. 3,00,000/-	29	41.1034	18.4359	
		Above Rs. 3,00,000/-	13	52.1538	7.3240	
5.	Nativity	Urban	101	43.6040	15.6685	2.745*
		Rural	85	48.1529	14.2741	
		Semi-urban	96	44.6979	15.7688	
		Tribal	18	52.4444	10.4893	
6.	Stream of Higher Education	Science / Engineering	108	47.8611	14.8011	1.602NS
		Commerce Management	97	44.6804	15.6901	
		Arts / Social Science	95	44.5158	15.0233	

** Significant at 0.01 level, * Significant at 0.05 level, NS Not Significant

From the table we can observe that the influence of existing institutional environment of ethical practices the opinion of the respondents

The institutional environment the responses of the students informs that female respondents obtained more men scores with SD of 15.067 than male students with mean score of 45.4093 and standard deviation is 15.2893, the calculated t value is 0.557 which is not significant. This means female students are satisfied with the institutional environment in ethical practices

The opinion in relation to social class indicates scheduled tribe students are better opined with a mean score of 51.2174 and obtained SD is 8.0054 and followed by forward caste (46.5310, SD 15.8391), backward caste (44.82000 SD 15.0783) the scheduled caste respondents obtained a low mean score of 43.9322 with SD value 15.9891. the obtained f value is not significant.

Regarding parental occupation background thee students from artisans families have better mean score of 51.4000 and SD value is 11.3314 followed by family background of non agricultural labour, agriculture labour, agriculture and business community. The student's parents working in government service have secured low mean score of 44.1753 and SD value is 15.5851 than their counterparts. The calculated f value is 0.989 is not significant.

About parental income the students with income of above 3 lacs per year agreed that institutional environment is good in ethical practices and the men score is 52.1538 and 7.3240 as SD value. Students with a family income of Rs. up to 3 lacs per year are getting low mean score of 41.1034 with 80.4359 SD value. The estimated f value is 4.365 and is significant at 0.01 levels.

Thee influence of nativity, the respondents belong to scheduled tribe caste have high mean score of 52.4444 with SD value of 10.4893 followed by rural, semi-urban and urban nativity students. The calculated f value is 2.745 which is significant at 0.05 level.

Regarding stream of higher education background respondents pursuing science/Engineering course obtained better mean score of 47.8611 and SD value is 14.8011 than their counterparts pursuing commerce and management (44.6804 and arts and Social Science 44.5158 the obtained f value is 1.602 and is not significant.

Table: 5.13: Impact of Factors -with reference to Ethical Practices - Socio-Economic Background of parents..

Mean impact on ethical practices pose, SD s obtained by the respondents belonging to different socio economic groups and the respective t/F values.

Sl. No.	Character	Variable	N	Mean	SD	T / F Value
1.	Gender	Male	193	48.7306	15.9426	1.132NS
		Female	107	50.9252	16.3400	
2.	Caste	Forward Castes	113	51.1858	17.2686	2.683*
		Backward Castes	105	47.0952	15.7685	
		Scheduled Castes	59	48.0508	15.7255	
		Scheduled Tribes	23	56.0870	8.9996	
3.	Occupation	Agriculture	65	50.6615	14.6557	3.174**
		Govt. Service	97	46.9278	15.0272	
		Business	104	48.1538	18.3392	
		Artisans	15	58.5333	10.5076	
		Agricultural Labour	11	57.4545	8.9372	
		Non Agricultural Labour	8	61.3750	12.2817	
4.	Income	Rs. 50,000/- below	39	54.3846	14.4106	4.688**
		Up to Rs. 1,00,000/-	112	45.6786	16.9367	
		Up to Rs. 2,00,000/-	107	52.9346	14.5513	
		Up to Rs. 3,00,000/-	29	44.3793	19.1897	
		Above Rs. 3,00,000/-	13	51.2308	5.1503	
5.	Nativity	Urban	101	46.3168	15.8505	4.223**
		Rural	85	51.4118	15.3555	
		Semi-urban	96	49.3021	16.5664	
		Tribal	18	59.6111	13.8315	
6.	Stream of Higher Education	Science / Engineering	108	51.0741	15.5411	0.769NS
		Commerce Management	97	48.5464	16.4785	
		Arts / Social Science	95	48.7263	16.3437	

** Significant at 0.01 level, * Significant at 0.05 level, NS Not Significant

Analysis of Data and Presentation of Report 165

The analysis was on the environmental conditions of the educational institution in relation to socio-economic and demographic variables of the respondents. In gender wise comparison female respondents obtained mean score of 50.9252 with SD value of 16.3400 better than male students and the 'f' value is not significant.

Rural social class background ST students have higher mean score of 56.0870 SD at 8.9996 shown forward casts 51.1858, SD at 17.2686 and other caste groups. Backward class responds has low mean score of 47.0952, SD at 15.7685 than their counterparts and 'f' value is 2.683, which is significant at 0.05 levels.

The opinion on institutional environment in relation to parental occupation as non-agricultural labour posses higher means score on 61.3750, SD at 12.2817 followed by artisans 58.5333, SD at 10.5076 agriculture labour posses' higher mean score on 57.4545 agriculture 50.6615. The business and Govt-service as occupation of parents, the students have obtained low mean scores of 48.1538 and 46.9278 and SD at 18.3392 and 15.0272 respectively. They obtained 'f' value 3.174, which is significant at 0.01 levels. Parental occupation has different impact on the attitude of students' responses as per the analysis.

In regard to family income and the opinion of the students on environment aspect, respondents belongs Rs. 50000/- per annum obtained higher mean score of 54.3846 than other income groups and calculated 'f' value is 4.688 and is significant at 0.01 levels. Income has no direct bearing on the opinion of respondents in environmental aspects. The respective values of SD are for students having the parental income less than Rs.50,000/- the SD is 14.4106, for less than Rs.1 lakh income group 16.9367, for students less than Rs. 2 lakhs the SD is14.5513, for students of income less than Rs.3 lakhs is 19.1897 and for the students hailing from the income group above Rs.3 lakhs the SD is 5.1503 respectively.

The influence of nativity of the respondents in assuming the institutional environment, the tribal respondents having more mean score of 59.6111, SD is 13.8315 followed by rural 51.4118, SD at 15.3555 semi-urban 49.3021, SD at 16.5664 and urban 46.3168, SD at 15.8505. Respondents practicing traditional customs have impressed with more favourable responses than pupil exposed to modern trends. The calculated 'f' value 4.223 is significant at 0.01 levels.

Regarding educational stream and student's opinion, respondents studying science/engineering obtained more mean score at 51.0741, SD at 15.5411 than their counterparts. The respondents pursuing commerce/management 48.5464, SD at 16.4785 and arts/social science 48.7263, SD at 16.3437 are obtained level mean scores. The 'f' value is 0.769, which is insignificant.

In the next chapter the brief summary of all the important concepts of all the previous chapters along with the conclusions, suggestions for further research were taken up.

"A Poem begins in delight and ends in wisdom."

Robert Frost.

Chapter – 6

Abstract

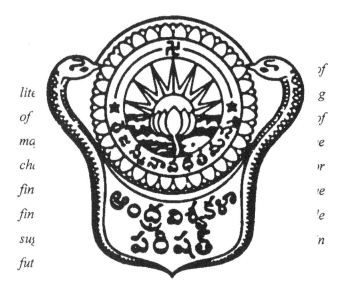

lite
of
ma
ch
fin
fin
su
fut

SUMMARY AND CONCLUSIONS

Summary and Conclusions

"Plants are shaped by cultivation and men by education. We are born weak, we need strength; we are born totally un-provided, we need aid; we are born stupid, we need judgement. Everything we do not have at our birth and which we need when we are grown is given us by education."

 Rousseau.

Every individual is born with some innate talents and capabilities and it's the duty of the civilized society to bring out the best and make him pursue his individual happiness by means of his talents and also be productive to the society in contributing to the common good and general happiness through his endeavour. The tool is always being the education system with all its ramifications and with right direction and right management tenets. In this process the educational system does have the potential to devise a right society which in turn will decide the quality of life, trade and commerce and becomes the main igniter for the wheel of development.

A dynamic society keeps on throwing challenges with the fleeting time and in changing environment with diverse problems in all aspects and education is no way different from this. In other words the quality and nature of education is being constantly being influenced by the changes in the ethical and environmental issues of the contemporary society and its management styles in delivering the goods. More pertinent is the higher education system as being the last leg of instruction before one enters into the life in a serious way in pursuit of individual happiness as well as social contribution through ones personal efforts and levels of productivity. Higher education results in better skills of workforce and results in better productivity that fosters development of a society and a nation as well. The relation between the education and economic development is synonymous and also determines with the social change and transformation into a better and civilized society. A well devised education system gives rise to increased gross national product (GNP) cultural richness, increased efficiency and effective governance. Off late in Indian educational system in general is fraught with maladies like resource crunch, credibility of institutions, commitment, efficacy of functioning and several other problems like moral degradation and poor management techniques.

India like any other country depends on the development of its educational sector especially the higher education to be competitive in the new world order of globalization which calls for nothing but efficiency and effectiveness in delivering the goods in different fields of development be it agriculture, science and technology, medicine, law, or the other related areas of development. In this process it is a fact that there is a serious dearth of skilled professionals and the educational system is unable to provide the number of quality hands even though sizable quantum of resources are allocated in this direction.

The chief areas of concern are.

1. Access to education is to be streamlined cutting the barriers of socio-economic, linguistic, geographical and other native maladies like caste etc.

2. Quality of education including infrastructural problems and the teacher and the processes quality

3. Resource allocation being very poor at 3.5% of GDP when the ministry of Human Resource Development recommends a 6%.

4. Department of higher education 2007 Government of India report reported that at elementary level 85%, at secondary level 39% and at tertiary stages of education a dismal 9% of gross enrolment ratios are recorded during the year 2003 - 04.

Philosophy - Ethics

Philosophy, is love for wisdom and wisdom being the essence of educational system, from ancient time to modern times the fundamental issues of ethics being, good, bad, right, wrong, justice, courage, happiness, freedom, benevolence, pity are all being treated as virtues in different degrees by most of the schools of thought.

Since the beginning of the civilization man is in constant pursuit of happiness in all his endeavours and different schools of philosophy have viewed and interpreted this in different dimensions as per their views. The central idea is that when there is a conflict between need and ideal it is always the ideal that is sacrificed for fulfilling the immediate need in pursuit of happiness.

Utilitarianism believes that an action is right when it is useful in promoting happiness and what the theory explains is the sum of pleasures. In other words pleasure is good and displeasure of pain is bad and also believed that an action is considered good and is better when it is capable of generating more pleasure than an action which generates less pleasure, as advocated by John Stuart Mill

Ethics of Immanuel Kant

Immanuel Kant (1724 -1804) was born in Konisberg, Germany. In spite of its various attractions Utilitarianism is able to give only a reasonably accurate picture of everyday moral judgements and is unable to meet the needs of philosophical theory and as a coherent system of ethics in a comprehensive way.

Kant distinguished the categorical imperative (moral) from Hypothetical (prudential or technical) Imperative. The first is concerned with the *form* from the categorical imperative the second is concerned with the *concept* and the third *links* these together. Kant's categorical imperative concerns the form and gives rise to the statements of universal moral judgements like 'ought to do' in nature.

In the second formulation that is the hypothetical imperative Kant substantiates or provides reason for 'right action' and *treats man as an end* but never as a means. To treat man as an end is to make his ends your own and regard his choices as your own and deal accordingly. This kind of treatment of ethical principles, by Kant has given rise to ethical concepts like democracy, justice, liberty, fraternity very strongly and subsequently called as an *ethics of democracy*.

Main trends of ethics in 20th century

Given the resources and constraints at the disposal of the investigator it is beyond the capacity to indulge in a detailed chronological depiction of the process of the ethical trends of 20th century. Accordingly the investigator attempted a modest and brief account of the dominant schools of ethics of contemporary times. In this process the review is confined to *ethics of*, irrationalism, formalism, and naturalism and axiological and theological schools rather than the subject matter of these schools.

Existentialist ethics

Existentialism unlike other philosophical schools places a special emphasis on morality. The entire philosophical literature is moral in nature. Schopenhauer recognized that the world was determined by the subject and also at the same time attached significance to will that exceeded the boundaries of human existence and assumed a cosmic meaning. Moral imperative is given an ontological status by existentialism. Morality as interpreted by existentialism that social morality is unreal and 'real' morality lies outside the society.

Pragmatism

Charles Pierce (1839-1914) is the first person to use the word 'Pragmatism' and this school of thought was developed mostly in1800 AD in United States of America. In pragmatism theory and practice are not separate spheres. John Dewey has put it that there is no question of theory versus practice but good practice versus bad practice. Pragmatism treats that there is no difference between facts and values. Both facts and values have cognitive content: knowledge is what we should believe; values are hypotheses about what is good in action. Pragmatist ethics is broadly humanist because it sees no ultimate test of morality beyond what matters for us as human beings. Good values are those for which we have good reasons.

Ethics: A Perspective

As ethical practices too play vital role in promoting and strengthening higher education, an attempt is made to understand the basic issues related to Ethics. Many people tend to equate ethics with their feelings. However, being ethical is clearly not a matter of following one's feelings. A person following his or her feelings may recoil from doing what is right. In fact, feelings frequently deviate from what is ethical. It is applicable to an atheist a materialist in other words to all the rational human beings. Being ethical is also not the same as following the law. The law often incorporates ethical standards to which most citizens subscribe.

Finally, being ethical is not the same as doing "whatever society accepts." In any society, most people accept standards that are, not harmful, that are convenient and that fosters happiness individually in the beginning and ultimately collective happiness or majority happiness.

First, ethics refers to well based standards of right and wrong that prescribe what humans ought to do, usually in terms of rights, obligations, benefits to society, fairness, or specific virtues. Ethics, for example, refers to those standards that impose the reasonable obligations to refrain.

And, ethical standards include standards relating to rights, such as the right to life, the right to freedom from injury, and the right to privacy. Such standards are adequate standards of ethics because they are supported by consistent and well founded reasons.

Secondly, ethics refers to the study and development of one's ethical standards like feelings, laws, and social norms. Ethics is concerned with what is right or wrong, good or bad, fair or unfair, responsible or irresponsible, obligatory or permissible, praiseworthy or blameworthy. It is associated with guilt, shame, indignation, resentment, empathy, compassion, and care.

It is interested in character as well as conduct. It addresses matters of public policy as well as more personal matters. On the one hand, it draws strength from our social environment, established practices, law, religion, and individual conscience. On the other hand, it critically assesses each of these sources of strength.

Socrates asks Euthyphro to define justice when he is set to punish his father (not knowing what is justice) for killing his servant and in this process he tries to elicit an answer for justice through his dialectic method he asks that '**what all just acts have in common that makes them just**'. Here what Socrates demands is a definition that captures the *essence* of justice in all of its instances, probably holds good even today. In this process the act that generates maximum happiness to majority of the people is being considered as ethical, moral, good, virtue and the opposite dimension is considered as unethical, immoral, bad and vice.

18th century philosopher Thomas Reid compares a system of morals to "laws of motion in the natural world, which, though few and simple, serve to regulate an infinite variety of operations throughout the universe."

Ethics and Childhood

A new born child on his birth is born with some kind of knowledge and subsequently the sum total of interactions one have with parents, relatives peers, friends, teachers will let the kid to change his perception from time to time. And this change in perception of the individual from with reference to a particular instance is what it is called and defined as learning. (Hillgard Atkinson and Atkinson).

Here psychologists provide the basic stuff that ethics at the time of birth is something and as you keep on growing with the change you keep on adding different sets of rules as does and don'ts which in other words is called as ethics. And this is the line of thinking that is considered as virtue that is useful for creating maximum happiness with minimum damage.

Children's introduction to ethics, or morality, comes rather early and has a fair degree of moral sophistication by the time they enter school. What comes next is a gradual enlargement and refinement of basic moral concepts, a process.

As philosopher Gareth Matthews puts it moral development is ... enlarging the stock of paradigms for each moral kind; developing better and better definitions of whatever it is.

Psychologists such as Jean Piaget (1896-1980), Lawrence Kohlberg(1927-1987), Carol Gilligan(1936), James Rest(1932-1999), and many others provides strong evidence that, important as feelings are, moral reasoning is a fundamental part of morality as well.

Piaget and Kohlberg established that there are significant parallels between the cognitive development of children and their moral development.

Kohlberg's account of moral development has attracted a very large following among educators, as well as a growing number of critics. He characterizes development in terms of an invariable sequence of six stages. The first two stages are highly self-interested and self-entered. Stage one is dominated and the next two stages are what Kohlberg calls conventional morality. Stage three rests on the approval and disapproval of friends and peers. Stage four appeals to "law and order" as necessary for social cohesion and order. Only the last two stages embrace what Kohlberg calls critical, or post-conventional, morality.

Descriptive and Normative Inquiry

Ethics normally has got the concepts some we feel as we know and understand thoroughly and some we are not sure and sceptic about. It is in other words the subjective feeling which is subject to acceptance by one and all when it comes to universal acceptance of the feeling which may be called ethical, moral etc.

Here one has to understand that there are two stages, i.e., subjective feeling – the first stage and acceptability universally - the second stage.

Psychologists, sociologists, and anthropologists endorse the values that people actually accept as values they *ought* to accept. To ask what values people *ought* to accept is to ask a *normative*, rather than simply a *descriptive* question.

iii. Philosophical Ethics

Traditionally, ethics has been dealt in philosophy and it examines basic questions about what our values should be, what, if any, fundamental grounding they can be given, and whether they can be organized into a comprehensive, coherent theory.

Thomas Reid opines that we do not have to be philosophers in order to think philosophically and we do not need to be a Plato or Aristotle in order to know our way about morally. He is also telling philosophers that in framing their theories they need to respect the understanding that of ordinary people.

Aristotle's account of the virtues, Immanuel Kant's categorical imperative and John Stuart Mill's utilitarian theories, all begin with what they take to be commonly accepted moral views

iv. Common Moral Values

Philosopher Sissela Bok states that we share some basic values and our desire to *get to the bottom of things* often blocks gaining a clearer understanding of what we have in common and notes that we may feel we need a common base from which to proceed. But there are different ways in which we might express what we think we need. Bok mentions ten different ways which detailed in the earlier chapter.

She states that positive duties are like, mutual support, loyalty, and reciprocity; negative duties to refrain are like harming others; and norms for basic procedures and standards for resolving issues of justice. And also states that these values are necessary for human coexistence at every level-in one's personal and working life, in one's family, community, and nation, and even in international relations. She states that when the above tenets are followed one will have positive and not followed will have negative consequences both global in dimension.

Kant opines that it is inappropriate to treat persons merely as means to the ends of research. As regards justice, it is generally agreed that discrimination in the selection of research subjects is inappropriate and that special attention needs to be given to especially vulnerable groups such as prisoners, children, and the elderly.

c. **Environment for Developing Higher Education**

To sum up, ethics has omnipresence in human life and so do its influence on higher education. Though ethical practices have its impact on Higher education in India yet there are other factors which also have an impact on the same and these factors are a conglomeration of various issues drawn from social, economic, ethics, values, traditions, culture and other related aspects.

Prior to independence, the growth of institutions of higher education in India was very slow and diversification in areas of studies was very limited and after independence, the number of institutions has increased significantly. With the increase in the number of educational institutions the need to evolve effective methods to handle higher education also has improved. Because learning is a complex process and the factors affecting the achievements of pupils can be broadly categorized into school-related and household related.

Children who are born to educated parents and competent teachers do very well, and are able to find jobs demanding high productivity but the average student with poor skills, results in massive unemployment even after several years of schooling, or even college education. Therefore, increasingly it is being realized that only by improving the quality of education can the positive effects of growing enrolments be sustained.

The Public Report on Basic Education (PROBE) says, "Quality education", however involves certain minimal requirements such as adequate facilities, responsible teachers, an active class room and an engaging curriculum.

India, believes through education only the complete human personality will come to be emphasized more and more imperatively. India also visualizes that contemporary problems can be resolved only if human nature is so changed that mutual goodwill and spontaneous drive to cooperation become ingrained in the human consciousness. India, therefore, visualizes a number of tasks that relate to the creation of a new society that is non-exploitative and non-violent in character by virtue of the integrated personalities of the constituents through higher education.

The educational environment which influences the Higher education can be considered in a three dimension way. The first dimension is the student itself. How the student conceives the teaching methods, teachers, attitude of parents etc.

The other dimension is the issue of the social and physical atmosphere prevailing in the education institution wherein the student is pursuing his interests in Higher education. The third dimension is the parents of the students itself. It is the influence of parents on the student matters most.

The new vision of contemporary higher education is looking for the following objectives are being emphasised: Education aims at liberation, education, being an evolutionary force that enables both the individual and the collectively to evolve various faculties and to integrate them by the superior intellectual, ethical, aesthetic and spiritual powers, education should be developed as a harmonising force, and education should be so designed carry the heritage and aim at transmitting knowledge to the new generations.

Thus, the process of education, especially Higher education is a matter of complex issue wherein certain ethical practices, parental influence and the education institution itself play a concurrent role in promoting the same. Hence, an attempt was made through this research study to understand the ethical practices being followed by the students, parents, and the effect of environment in educational institutions as well as students while pursuing higher education.

Management a perspective

Management of higher education in India is being viewed as a powerful tool for fostering the development in the recent years by the rulers and administrators. Management also plays an important role in judiciously administering the valuable and meagre resources and help obtaining the desired results in achieving the nation's goals. In order to provide an understanding on the basic tenets of Management, a brief review of Management Thought and its transformation over years is attempted in nutshell.

Evolution of Management:

The concept of management and its practice can be traced back to 3000 B.C to the first government organization developed by the Sumerians and Egyptians. The early study of management as we call it today is 'Classical Perspective' of Management. In the following few pages a brief review of management and its development and thought is attempted to by the researcher.

Management - A perspective:

Frederick Winslow Taylor (1856 -1915)

Classical perspective of Management has stressed the need for rational and scientific approach to study and problem solving and the approach to pinpoint organizational plans, tools, structures, systems, jobs, employee's roles and strict adherence to objectives/goals achievement became the basic pre requisite.

Frederick Winslow Taylor (1856 -1915) preached for deviating from thumb rules to precise measurement of procedures for various aspects of work and related activities. He stated that 'in the past man has been the first but in the future the system must be first'.

Henry Gantt L:

Scientific Management:

Henry Gantt has developed bar graph that measures planned and completed work, along each stage of production by time frame or time elapsed.

Frank B Gilbreth & Lillian M Gilbreth (1868 -1924) pioneered time and motion study, i.e., the former has divided work into various movements and the later provided the scientific measurement.

Max Weber (1864 - 1920):

Max Weber introduced the most of the concepts on bureaucracy. In this regard it is worth noting that there existed a hierarchy in the organizations and the individuals and individual goals attained more importance than the group or organizational goals.

Mary Parker Follett (1868 - 1933) was a major contributor for the administrative principles' with emphasis on worker participation and shared goals among managers.

'General and Industrial Management' of Henry Fayol discussed 14 general principles of management like, 'Unity of command - one superior or boss,' 'Division of work - workers produce more because of specialization,' 'Unity of Direction -

similar jobs grouped together' and 'Scalar Chain - a chain of authority that starts from the top to bottom of the organization'.

Chester I Barnard (1868 - 1961) has contributed two important concepts in management i.e., informal organization and the acceptance theory of authority which state that people are not mere machines and people follow orders as they perceive positive benefits by doing so.

Behavioural Sciences Approach-Humanistic Perspective - Human relations movement:

Drawing heavily from the principles of behavioural sciences is the concept of 'Organization Development,' like human relations, team spirit etc. but not money proved to be the motivators. This non monitory concepts scoring over monitory benefits is called as Hawthorne effect. This humanistic approach for management vouched for designing the jobs so as to bring out the best from the workers and to utilize their full potential. The advocates of this movement were Abraham Maslow (1908 -1970) who proposed his hierarchy of needs theory from physiological needs to self actualization being the highest need.

Doglus McGregor (1906 -1964) proposed his famous X theory and Y theory, wherein X theory views workers with negative assumptions that workers detest work and to be forced to work and to be closely supervised, etc. and Y theory being with positive assumptions that workers like work and enjoy it and keep on producing without any supervision with self motivation.

Management Science Perspective:

Management Science Perspective utilizes operations research, operations management, and information technology and the connected sub branches of knowledge like mathematical models, forecasting methods, linier and non-linier programming, queuing theory, scheduling, simulation, Enterprise Resource Planning, e-commerce etc .for improving productivity.

Recent Trends in Management:

Management always preaches 'The Right Way' in approach of the problem and finding the solutions by utilizing the optimum of the resources and getting always the best in any given situation, and education being no exception.

Systems Theory

Systems concept has broad meaning and systems will have boundaries and they also interact with environment and the organization In this aspect management treats organizations as a set of systems depending upon the area of operation and the nature of the production activity be it a product or service which are inter connected in achieving the goal or objective.

Mathematical or Management Science Approach

The methods of mathematics, inductive, deductive and symbolic logical principles are utilized wherever they are applicable to find solutions for problems in management. Even though many problems can be modelled on mathematical basis there is a serious constraint and is not feasible always. This approach handles the problems in cause and effect and contingency aspect.

Roles Approach of Mintzberg

He opined that managers have three basic and fundamental roles i.e., 1) interpersonal 2) informational and 3) decision roles. The limitations for this system are the sample taken for study was very small based on which the conclusions are drawn.

7-S framework approach of Mckinsey

Mackinsey has postulated an approach using seven concepts all starting with the letter 'S' and making a management model to explain its role. They are 1. Strategy, 2. Structure, 3.Systems 4 .Style 5.Staff 6.Shared Values and 7.Skills.

Operational Approach

This approach of the management system is based on managerial functions viz. planning, organizing, staffing, leading and controlling. However the major weakness of this approach happens to be that of not identifying or representing the vital function of 'coordination' and its purpose in management.

Higher Education

Higher education in India is one of the most developed in the world by means of its sheer spread and magnitude in catering to the needs of innumerable numbers. There are about 227 government recognized out of which 20 are central Universities, 109 are deemed universities and 11 are open universities.

Institutions governing Higher Education:

The institutions governing higher education in India are independent and multifaceted. University Grants Commission (UGC) is responsible for coordinating and maintaining standards of Higher Education. In addition to UGC professional councils are responsible for recognition of courses and promotion of professional institutions for providing grants and determining other academic aspects. A list of twelve (12) institutes was given in the relevant chapter.

Central government is responsible for major policy decisions through UGC under establishments, and state governments are responsible for establishment and up keep of State Universities. The coordination between centre and state is ensured by Central Advisory Board of Education (CABE).

Academic Qualification Framework – Structure

There are three principle levels of qualifications in India in Higher Education after completion of twelve years of schooling. These are:

- Bachelor / Undergraduate level.
- Master's / Post-graduate level.
- Doctoral / Pre – doctoral level.

Autonomous colleges

As per national policy of education 1986, a scheme of autonomous colleges 138 in India have come into existence. The nature is that they are free to decide on their own curriculum and conduct their examination/evaluate and award the degrees. They also can develop and propose new courses for approval for the university.

Accreditation

University Grants Commission in the year 1994 has established an autonomous institution and invested with the responsibility of accreditation called 'National Assessment and Accreditation Council (NAAC). For this purpose UGC has prescribed certain norms and processes in this regard.

Open University System

With the advent of information technology and subsequent information revolution in the year 1982 Andhra Pradesh started 'Andhra Pradesh Open University' and in the year 1985 government of India has stated another university Indira Gandhi National Open University (IGNOU). The modus opherendi is elaborately dealt with in the relevant chapter earlier.

Important Institutions of Higher Education

University Grants Commission (UGC) established in the year 1956 discharges the duty of determination and maintenance of standards of teaching, examination and research in higher education.

All India Council for Technical Education (AICTE), looks after all the major Engineering, Pharmacy and MBA colleges are affiliated with. It formulates, plans, and maintains norms and standards and assures and ensures quality, priority of education. The detailed account of operations of AICTE was given in the relevant chapter earlier.

The mile stones in the history of Indian Higher Education like the establishment of different Indian Institute of Technology's (IIT), Indian Institute of Management's and other relevant bodies/organizations which aid and administer higher education in India was furnished in the relevant chapter earlier.

Challenges before Indian Education

Coming to the challenges that are faced by the Indian education there are many and the main challenges are, Access, Participation, Quality, Relevance, Management and Resources.

In conclusion of this part, i.e. higher education it is important to note that the present less of resource allocation of 3.5% of GDP in the year 2004-05 is falling short of the requirement of the resources which got to be to the tune of 6% of the GDP.

REVIEW OF LITERATURE

Review of literature covers both Indian foreign studies and accordingly the summary of the studies is furnished below.

John Best (1982) opines that the review of literature reflects the understanding of the researcher of the problem and helps avoiding duplication of effort in the same direction and on similar topics.

Summary and Conclusions						188

Clark (1927) found that students of graduate parents had ranked higher in scholarship examinations.

Austin (1924) identified the relation between father's occupation and dropouts in college education.

Sinha, (1970) Wig and Nagpal (1970) identified relationship of low achievement of students vis-à-vis occupational categories like agriculture and business, found positive.

Griffiths (1926) established a close positive relation between family size and superior school grades.

Havighrust (1964) identified enriched socio-economic status is correlated positively with one hear above the actual grade level in 21 Chicago district schools.

Mishra, Dash and Pahdi (1960) identified correlation between home environment and IQ test scores and school achievement.

Menon (1973) and Anand (1973) established relationship between socio-economic status and academic achievement.

Abraham (1974) and Basavayya (1974) identified relation between socio-economic status and English language proficiency and parental occupation.

Bahaduri (1971) identified higher socio economic status resulting in under achievement in school education.

Socio economic conditions contribute to academic achievement was identified by prakash Chandra (1975) ,

Homchandhuri (1980) , Khanna (1980) , Shukla (1984), Meharotra (1986), Mishra (1986), Singh (1986), Rathaiah and Rao (1997).

Sarah (1983) observed a positive co-relation between socio economic status and pupils attitude towards science education.

Das (1960) and Gopalacharyulu (1984) observed socio-economic status influence total academic achievement along with caste of students and teachers.

Pandey(1981) and Puri (1984) identified that urban environment fosters general academic achievement.

Paul (1986) identified parents aspirations and attitude had positive significant correlation in students academic achievements.

Gaur (1982) demonstrated birth order did not influence academic performance of siblings.

Lal (1984) has identified that protected attitude of parents positively related to the academic success of the boys.

Jagannadhan's (1985) study identified the positive significant effect of home environment on academic performance.

Trivedi (1987) correlated parental attitude and academic achievement positively.

Deve and Deve (1971) observed that the size of the family was not related to academic achievement and high parental income and high performance.

Dhami (1974) in his study has found that though the relationship between socio-economic status is co-related with academic achievement.

Srivatsva (1981) has identified that high parental income resulted in better academic performance of the wards.

Nemzek (1940) reported that high ability group children, are more from service men parents rather than businessmen parents.

Deshpandy (1984), established that there is no specific trend of organizational climate influencing students' performance.

Upadhyaya (1982) found that three aspects i.e. classroom environment, interpersonal relationships, goal orientation and system maintenance, was significantly related to academic achievement.

Role of society - Policy issues and Programmes:

The role of society is that creation of an environment with forward and backward linkages so that every student not only pursues his education and benefits with it.

Lunngdim (2000) observed that students face lots of hardship and also identified the need for counseling student on a constant basis to channalize their aspirations.

V.K. Patil (2000) observes that vocationalization of education is a bold initiative and it offers a new direction and to higher education to meet the global challenges in the new world order.

Summary and Conclusions 191

Dr. V. Kulandiswamy wants organizations like UGC are to be remodeled to meet the demands of the higher education system.

V.T.Patil and Dr Narayana stresses the need for 'External Quality Assurance" to maintain sustainability of colleges and universities in India.

Ashokan and Virk (2004) have highlighted the need for revamping the higher education system to the needs of students by UGC.

Powar K.B (2000) proposes that the divergence of educational system with all its constituent diversities is to be brought into one matrix to achieve synergy. He also opines that the distance education with its activity is able to attain 'access and equity'.

Higher education is oriented and influenced by the environmental demands of new world order i.e. globalization and market demands on the one side and the students desires to fit into the demands through which they can eke out their livelihood.

V.C. Kulandi Swamy (2006) observes that there is very little that has been done in the direction of higher education in India to build a solid framework for human resource development in India which is the need of the hour.

It is also observed that the poor preparedness of India in international competition in the new world order of opening up economy and education, for the foreign countries as per General Agreement on Trade in Services (GATS) will leave the country struggling to compete with.

Mohan Ram (2004), opines that "include models of thinking and feeling relevant to common life" and also has to specify the goals of education in general and that of the higher education specifically.

Niyati Bhat (2004) observes and calls for absolute diligence in determination of the policy matters and subsequent decisions regarding higher education..

Madan .V.D (2002) advocates the importance of technology implementation, in higher education to satisfy the needs of the stake holders especially the students and their end needs.

Venkataiah .S (2001) opines that in higher education the importance of effective use of technology in teaching and learning for the overall success of the institutions.

Oza N.B and Joshi.K.M (2001) advocated the implementation of management techniques like Total Quality Management (TQM). to utilize the latest developments of technology and industry and utilize them in their day to day problem solving with optimization of resources.

Bharat B Dhar (2008) has stressed the need for emulating the technological advancements in the field of information technology, strategies in internationalization of education in higher education along with research enabling to register patents with WTO.

V.K.Rao opined that higher education is to be administered with regular periodic reviews of policies and plans with reference to their consequences.

P.K. Nayak opines that higher education had to cater to the needs of variations exclusively different to local contexts at micro level.

Research Methodology

Research Objectives

The objectives of the research are furnished below:

1. To understand the social and economic profile of students participating in higher education.

2. To assess the perception and preferences of various students in higher education in their institutions about the ethical and management practices followed.

3. To assess organisational and managerial issues and problems encountered by promoters of higher education.

4. To assess the understanding and analyse opinion of students on the moral values protected and promoted by the educational institutions among stakeholders in higher education.

5. Suggest measures for better ethical and management practices in higher education.

IV. Research Methodology

i. Study Area

The study is conducted in Visakhapatnam district in Andhra Pradesh, India. The district is specifically selected since Visakhapatnam is the seat of higher education.

Studying the ethical, environment and management practices in higher education in Visakhapatnam would certainly provide better insight and hence the selection is made.

ii. Selection of Sample (Respondents)

Keeping the resources available with the researcher, a sample size of 300 students was selected for the purpose of the study. The sampling was done on the basis of random sampling method to avoid prejudice in selection of respondents for the study.

Study Area: A Description.

Visakhapatnam district lies on the northern coastal area of Andhra Pradesh in due course it was attached with Andhra Pradesh with the new reorganization of states under the Indian Union. Coming to physical features, a strip of plain land along the east coast followed by Eastern Ghats with hills, and in the North and west agency area or the forest area. The detailed description of the area was furnished in the relevant chapter earlier.

Major findings of the study:

Students – Response:

As described in the research methodology adopted for the study, a sample of 300 students was selected who were pursuing higher education in Visakhapatnam city. Further, in order to understand the impact of stream of education i.e. specific branches of education pursued, a sample of 100 students from each of the distinct streams of education were selected. The findings are listed below.

Socio-economic particulars

Community and Gender Profile of Students

- The analysed data in this regard as observed from the data presented, majority of them one hundred and ninety three (193) students (64.33%) were male and female represent the remaining.

- In regard to community profile, almost equal numbers of students were from backward community 105 students (35%) as well as other castes or Forward Caste students 113 (37.66%) community. Among the remaining students, communities of Scheduled Castes 59 (19.67%) and Scheduled Tribes 23 (7.67%) were quite dominant.

iii. Occupational Profile

- It is observed from the data presented, the occupational background of the students' families is dominated by Business (34.67%) and Government Service (32.33%). In fact, two thirds of the students (67%) were having the occupational background of either business or government service. The next major occupation prevailing among the students is agriculture (21.67%) and then followed in a nominal way in reference to artisans (5%), agriculture labour (3.67%) and non agriculture labour (2.66%).

- With reference to Arts and Social Sciences, the major occupational background is dominated by agriculture when compared to other streams of education, and students from the occupational background of government service and business were present more under the streams of science and technology as well as commerce and management when compared to arts and social sciences.

- In this context too, it may be pointed out that those involved in government service and business are in a better position to send their children for higher education in science and technology as well as commerce and management.

iv. **Profile of Annual Income**

- Among the 300 students selected for the study, slightly higher than one third of students (37.33%) were having an annual income up to Rs 1 lakh per annum and then closely followed by 107 students (35.67%) who reported having an annual income up to Rs 3 lakhs. Almost one tenth of the students (9.67%) reported that their annual income is up to Rs 5 lakhs and only 13 students (4.33%) reported that they were having an annual income higher than Rs 5 lakhs. The remaining 39 students were having an annual income up to Rs 50,000/- who mostly belongs to poverty group.

- Further in reference to different streams of education, it may be observed that students having an annual income up to Rs 1 lakh were mostly found in arts and social sciences stream, and similarly, higher income families naturally prefer education for their children in streams like science and technology as well as commerce and management.

- However, it may be pointed out that government is providing assistance as well as certain benefits to poor people for their children education and this in background, though substantial expenditure is involved in higher education, poor families could able to send their children to higher education.

Summary and Conclusions

v. Students' Place of origin

a. In terms of ethical practices and while going through the educational environment, the place of students' origin also matters most and from the data presented that most of the students were having either urban (33.67%) or semi-urban (32%) background. The remaining students were from rural areas (28.33%) in a substantial manner and very few of them (6%) from tribal areas.

b. Further, on close observation of students pursuing different streams of higher education, it may be seen that substantial number of students from the rural areas were found pursing arts and social sciences courses rather than the other streams of higher education.

c. On the other hand, the urban and semi-urban students were found more in the streams of engineering and technology as well as commerce and management.

vi. Distinction held by students in eligible examination

- It may be observed from the data presented that one fifth of the students 60 (20%) were reported scoring more than 90% in their eligible examination concerned. However, the distribution of sample in this regard is not equal when different streams of higher education were considered. Among these students, most of them were from science and technology stream (29 students) and then followed by commerce and management (22 students). Only 9 students were found under the stream of arts and science. Thus, those held highest bracket of marks were opting for higher education streams in the fields of science, engineering, commerce and management.

- Similar is the distribution of sample in reference to students who secured marks between 80% - 90% (23%).

- At the overall sample level, majority of the students obtained marks in the range of 70%-80% (37.33%) and that most of these students were distributed in arts and social sciences stream.

- To sum up, it may be pointed out that the students who secured higher marks in their qualifying examinations were mostly found in science and engineering stream and then followed by commerce and management. Students who secured comparatively lesser marks were found in arts and social sciences stream.

- It is observed that the sex-ratio among the sample of students has been in tune with the representation of boys and girls in higher education system.

- The occupational background of the students' families is dominated by government service as well as business and then followed by agriculture. The middle income-group of annual income (Rs 1 lakh to Rs 3 lakhs) is more dominant as well as lower income level between Rs 50,000 - Rs 1 lakh.

- It is observed that those who have scored higher marks in the qualifying examinations concerned were found distributed more in science and engineering stream and then followed by commerce and management and arts and social sciences stream of students.

- Thus, the sample of the study is in close proximity with the conditions and trends prevailing in the society and thus represents the general scenario.

- The investigator enquired about the respondents about their awareness on ethical practices. Out of 300 respondents only 133 (44.3%) have agreed that they have knowledge on ethical practices. But 167 (45.6%) respondents have agreed that they are ignorant about this. This indicates even the students studying at degree level have little or no reasonable level of knowledge about ethical practices.

Opinion on importance of ethical practices in higher education.

- On the need of ethical practices and their importance the response from the statement that it is essential to keep a seat of higher education with certain principles majority 227(75.7%) agreed.

- Regarding the statement 'does your institution promises to have such atmosphere' only 41 (13.7%) positively admitted but majority 212 (70.7%) have expressed negative opinion with the statement.

- The respondents have more positively inclined towards process of education in the institutions and 145(48.4%) preferred this method.

- Satisfaction towards the practice of moral and educational practices in, their institutions 154 (51.4%) respondents admitted that they are dissatisfied with the existing practices.

- Majority of the respondents 120(36.7%) feels that ethical practices in educational institutions are an asset for building a strong value system but nearly equal number of respondents 108(36%) have negative opinion.

- With regard to respondents satisfaction level about the practicing ethical values in the institutions majority 182 (51%) have expressed their satisfaction.

- Majority of the students 193(60.3%) have expressed positive ness that the role of the students is important.

- In view of the statement faculty's adherence to ethical practices majority 176(58.7%) have negative opinion and only 109 (36.3%) either strongly agreed or agreed in this regard.

- Majority 146 (48.7%) of the respondents have positive ness towards attention of the management in creating positive environment towards ethical practices.

- In regarding maintenance of ethical practices in their study environment 142(47.3%) strongly agreed and 76(25.3%) agreed and only 34(14.7%) have negated with the statement.

- A marginal majority of respondents' 155 (51.7%) feels that ethical practices in the institutions are the equal responsibility of management and students. But 121(40.3%) have strongly disagreed about equal responsibility.

- The responses indicate that 94 (31.3%) strongly agreed to this aspect helping them for their future career but majority 133(44.3%) disagreed with this statement. This indicates the respondents are not fully dependent on ethical practices and they have other plans also in planning their future career.

- Majority of the students 140(46.7%) strongly agreed that the influence of others have no impact if the person have self determination to adhere to ethical practices. Simultaneously 67(22.3%) respondents have an opinion that influence plays a role in the institutions.

Summary and Conclusions 203

- Majority 192(64%) are either strongly or agreed that the ethical practices will be innate and intrinsic in character rather than induced one.

- Similarly regarding statement of ethical practices is a trait which comes from the family background, majority 167(57.7%) expressed their acceptance with the statement and other respondents 80(26.7%) have negative assumption.

- As regards the social background affecting the adherence to ethical practices 163 (54.3%) expressed their agreement with different degrees in the positive and affirmed it. Thirty nine respondents 39(13%) have expressed that they have no idea and 98(32.7%) affirmed in the negative that the economic background does not affect their adherence to the ethical practices.

- Majority 190(63.3%) students agreed that the credibility and reputation of the educational institutions enhanced only on the ethical practices adopted by the administration rather than student's academic performance.

- Regarding student's opinion that their academic performance is to a depended upon the way their adherence on ethical practices, in this aspect majority 204(68%) either strongly agreed or agreed with the statement in the affirmative. Only 72(24%) disagreed that personal practice if ethical values have no relation to his academic performance.

- Majority 163(54.3%) expressed positive opinion that their social and economic background had an impact on their grooming attitude to ethical practices, only 12(4%) negated with this statement.

- The analysis was made to estimate the impact of institutional environment on the students to adhere to ethical practices.

- Majority 188(62.6%) of the respondents either strongly agreed or agreed that different social and economic backgrounds of respondents are not congenial to follow ethical practices as they keep on preaching different lifestyles and different beliefs and attitudes. 26(8.7%) respondents expressed that they neither agree nor disagree and 86(28.7%) are negative to the statement.

- Regarding the commercial environment in the institutions always a distraction and divert the strong intention to adhere to ethical practices majority of the respondents 160(50.4%) agreed with the statement and affirmed that commercial environment of managements is a distraction.

- With the statement that the ethical practices are sacrificed because of the commercial attitude of educational institutions 165(54.7%) either strongly

Summary and Conclusions

agreed or agreed and only 79 (26.3%) respondents strongly disagreed with this opinion.

- Regarding institutional physical environment is not conducive to the ethical practices 154 (51.3%) have positive opinion, but 113 (27.6%) disagree with their co students opinion.

- Similarly majority 143 (47.7%) respondents feel that academic atmosphere in their institution is not conducive to ethical practices and about 90(30%) expressed that the atmosphere helps them to adhere to ethical practices.

- Majority 178 (52.1%) respondents expressed their positiveness that the cordial relations among students made a way to ensuring adherence to ethical practices in institutional environment, only 51 (20.3%) respondents have negativeness in this regard.

- The impact on students to adhere to ethical practices only 125 (41.7%) strongly agreed that positive relation with faculty yield better results to follow ethical practices during their period of study in the institution.

- The physical surroundings play a vital role on ensuring ethical practices majority 195 (65%) have expressed positive opinion.

- The respondents 149 (49.7%) have an opinion that for effective ethical practices in educational institutions there must be a provision to establish educational counselling units.

- The efforts of management to inculcate ethical practices in the institution environment 98(32.7%) respondents strongly agreed and were positive on the efforts of their respective institutions.

- Two thirds 202(67.4%) expressed positive opinion that the educational background of parent have a greater influence on creating ethical practices among students.

- The influence of birth plan on inducing ethical practices, a marginal number of respondents 111(37%) disagreed with the statement.

- Nearly half of the respondents 149 (49.6%) agreed that in inducing ethical practices the family members education has a bearing.

- The respondents 90 (30%) disagreed with the statement, teaching staff is only responsible to induce ethical practices in the institutions.

- There is a mixed response from the statement that in spite of staff influence on students to create a right atmosphere on the part of the management is necessary, 153(51%) respondents have strong acceptance and the others have different views.

- In higher education while inculcating ethical practices, the practices adopted in previous level of education have an impact, a marginal majority 108(36%) disagreed with the statement.

- Majority 220(73.3%) agreed that the personal determination and discipline only matter in practicing ethical principles despite of influence of various factors.

- Majority 198(66%) respondents have an opinion that the stream/branch of education is more important than quality of educational institutions background in inducing ethical practices.

- The information on parental occupation on students' attitudes towards ethical practices, on analysis it is observed that the students from families belonging to fixed monthly regular income are lagging than business belongings to un-organized sector.

- The impact of family annual income on students shows students from business in score at surprisingly the students from business income groups have more positive opinion on ethical practices.

- The nativity plays a vital role to adjust to the environmental imbalances. From the analysis it was found that the more modern exposure on the part of the students will lead to change of attitude towards participating ethical values.

- The attitude of the students on ethical practices was calculated depending upon their discipline/faculty. It was observed that students from science and engineering discipline acquired higher mean score may be due to teaching learning environment of the class room.

- In nutshell, female students belong to artisans' families, higher income tribal relatively and studying science/engineering students having higher mean score than their counterparts.

Institutional Environment on Ethical Practices: in relation to the respondent's socio economic background.

Mean impact on Environmental conditions pose, SDs obtained by the respondents belonging to different socio economic groups and the respective t/F values.

- The analysis was on the environment conditions of the education institution in relation to socio-economic-demographic variables of the respondents.

- In gender wise comparison female respondents obtained score mean score of 50.9252 than male students the 'f' value is not significant.

- Rural social class background ST students have higher mean score of 56.0870 shown forward casts 51.1858 and other caste groups. Backward class responds has low mean score of 47.0952 than their counterparts and 'f' value is 2.683, which is significant at 0.05 levels.

- The opinion on institutional environment in relation to parental occupation as non-agricultural labour posses higher means score on 61.3750 followed by artisans 58.5333, agriculture labour posses' higher mean score on 57.4545 agriculture 50.6615. The business and Govt-service as occupation of parents, the students have obtained low mean scores of 48.1538 and 46.9278. They obtained 'f' value 3.174, which is significant at 0.01 levels. Parental occupation has definite impact on the attitude of students' responses as per the analysis.

- In regard to family income and the opinion of the students on environment aspect, respondents belongs Rs 50000 per annum obtained higher mean score of 54.3846 than other income groups and calculated 'f' value is 4.688 and is significant at 0.01 levels. Income has no direct bearing on the opinion of respondents in environmental aspects.

- The influence of nativity of the respondents in assuming the institutional environment, the tribal respondents having more mean score of 59.6111 fallowed by rural 51.4118, semi-urban 49.3021 and urban 43.3168. Respondents practicing traditional customs have impressed with more favorable responses than pupil exposed to modern trends. The calculated 'f' value 4.223 is significant at 0.01 levels.

- Regarding educational stream and student's opinion, respondents studying science/engineering obtained more mean score at 51.0741 than their counterparts. The respondents pursuing commerce/management 48.5464 and arts/social science 48.7263 are obtained level mean scores. The 'f' value is 0.769, which is insignificant.

Field Observations

During the period of data collection the investigator used to make observations which are of some significance for the research topic. In this process the following observations are recorded.

Managements and Faculty

- Managements have the problems like non receipt of funds from the government with reference to dues like reimbursement of fees and scholarships on time leading to financial constraints and are effecting the management in quality aspect of education which is to be sorted out at the larger interest of students of higher education.

- Management in another college is having the nature of diverting the funds that are collected as fees from students to other business ventures ignoring the salaries of the college staff both teaching and non teaching. This kind of attitude is to be sorted out by evolving a fool proof strategy. These aspects creates psychological disturbances and leads to apathy and disinterest among faculty to concentrate on other than academic activities, like behavioural patterns of students and in organising social and cultural activities.

- Management used to express that the investment on the college is always viewed as an investment in terms of real estate and they are mostly satisfied by the appreciation of the land and other infrastructure rather than building up academic excellence, in case of private management institutions. This trend is to be arrested so as to benefit the students as well as the managements in the long run for the benefit of the higher education.

- Managements are employing mostly one senior man in the department concerned with all credentials in faculty and the rest freshers who are just coming out of colleges who accept anything that is offered to them and treating the teaching as a stop gap arrangement with inadequate salaries and job insecurity. This matter is to be looked into so as to improve the quality of the teaching faculty.

- Faculty working in the affiliated colleges except in the case of university and most of them are not getting the pay and perks as per the stipulated norms in higher education, because of which students are not getting the best of faculty and are at loss which is to be rectified.

- Managements are unable to staff the departments with experienced faculty members who have the wherewithal to guide the students in academics and extra curricular activities.

Students

- Students are always under constant pressure to perform both in examination and after examination because of the fact that unless they get a job at the end of their career they are in a difficult position. This pressure is to be taken care of by means of strong academics and other capacity building measures like equipping the students with better communication skills and character building measures, by allocating some time in the curriculum itself.

- The rat race for employment after education is leading to a unhealthy competition in concentrating more on modes of job getting skills rather than the fundamentals of academics. This tendency is giving rise to delivery of poor quality students out of the institutions.

- Students mostly being adolescents at the time of admission into higher educational institutions are faced with the dilemma of deciding the right path towards their career because of lack of proper career and counselling opportunities in the college. If it is coming from the parents it is fine and those who don't get it are falling prey to deviant tendencies and are lagging behind in studies and unable to complete the course. This aspect is to be looked into and proper remedial measures are to be taken up.

- Students in most of the colleges are getting their instruction just to hand to mouth and when it comes to extra curricular activities it is even worse for want of proper play ground, facilities like library etc. which is also one of the important factors which has the potential to affect the environment of higher education.

- Another important aspect is that students even though proper facilities are provided the timings of the colleges are becoming very inconvenient for majority of the students as most of the students come and go to college by college busses which leave immediately after the college hours, depriving of any time for physical educational activities or for making use of library. A properly designed time table will help avoiding the above problems of the students which is to be looked into.

- Students used to come to the college late and usually are absent for the first hour and spend time in the canteen and in the campus loitering and chit chatting and the same set of student used to leave the college in the lunch break. A proper counselling and disciplinary mechanism is to be evolved to set this right for the benefit of the students.

- Lack of neither inspiring teachers nor leadership in the faculty because of their own problems of survival with the academic careers. From time to time the managements of the colleges have to endeavour to arrange for interaction with people having remarkable achievements in life in various fields of life.

Implications of the study

The higher education in India even though is diversified is unable to get the desired results in the quality of education and there is always a cribbing from the stake holders that there is a thorough degradation of morals and ethics in the new generation of students pursuing higher education.

In consequence the employability of the students is coming down and accordingly giving rise to diverse social dissonance resulting in the wastage of valuable energies of youth. Accordingly the money and time that is spent on higher education both by the parents, government, and the students is not finding proper utilization and not yielding results which are aimed at in other words the resources are wasted.

- The well designed education system is the need of the hour for new world order of global competition to deliver quality and well trained hands ready to use in employment immediately after study in various fields of activity. The various findings can help fine tuning the higher education system in India in delivering the goods for the stake holders of higher education.

- Poor curriculum and course study are unable to inculcate and create interest in the students in higher education. The need for a curriculum that can interact with the needs of the individual and the society is the need of the hour and can be designed by keeping the various factors under scrutiny for the benefit of the students and society.

- Improper capacity building measures on the part of the colleges are resulting in despondency, disinterest and despair among the students at the end of their higher education. As knowledge is power and being able to throw light on the intricacies of higher education vis-à-vis its goals is a job done well for the benefit of the society and the nation. In this process students can be benefited by implementation of the findings of this study.

- Except a few, majority of educational institutions are unable to fetch a livelihood for the students immediately after the end of the academic programme. A well designed educational programme will ensure the expected results for all the stake holders and ultimately yields fruits. In view of this a practical curriculum is to be designed from time to time for the students of higher education enabling the demands of society reflects in education system.

- The stipulated goals and are not attained successfully in higher education in India.

- 'A sound mind in a sound body' is the age old adage and in line with this the higher education may be designed with all its ethical tenets to help all the stakeholders of higher education.

Suggestions for further Research:

1. The management practices adopted by secondary education, 10+2 level in aided institutions, unaided

institutions and religious institutions can be taken up both at macro and micro levels.

2. A study on the effect of ethics and environment of boys and girls in higher education management and its can be taken up separately for better understanding of the problem closely.

3. A study on the ideal environmental settings for higher education in the fields of Engineering, Management and Arts and Social Sciences can be taken up so as to find out the true attitudes of the students which may help planning higher education more effectively.

4. A study on the effectiveness of higher education vis-à-vis employment in the fields of Engineering, Management and Arts and Social Sciences will throw light for better planning of higher education.

5. Micro-level studies can to be taken up to estimate the relation between discipline and ethical environment of the educational institutions at different levels of education like primary, secondary etc.

"The end of all knowledge must be building up of character"

_____ Mahatma Gandhi.

Abraham Harold Maslow (1943) *A Theory of Human Motivation* (originally published in *Psychological Review*, 1943, Vol. 50 #4, pp. 370-396).

Abraham. M (1974): "Some factors relating to under achievement in English of secondary school pupils" Ph.d. thesis.

Adorno, Theodor W. (1989). *Kierkegaard: Construction of the Aesthetic*. Minneapolis: University of Minnesota Press.

Agarwal, P, "Higher education in India: The Need for a Change" Indian Council for Research on International Economic Relations, 2006.

Agarwal, Pawan. "Higher Calculation in India: The need for Change".Oxford University Press, (2006).

Albert Camus, The Artist in the Arena (1965), by Emmett Parker.

Alexander, Thomas. *John Dewey's Theory of Art, Experience, and Nature* (1987) SUNY Press.

Amutabi. M. N. & Oketch .M. O, 'Experimenting The Distance Education: The African Virtual University (AVU) and The Paradox of The World Bank in Kenya', International Journal of Education Development, 23(1), 57-73,(2003).

Anand, CL (1973), "A study of the effect of socio economic environment and medium of instruction on the mental abilities and the academic achievement".

Aristotle, Renford Bambrough (1963) "The philosophy of Aristotle", New American Library, 1963O riginal from the University of Michigan.

Atwell, John. *Schopenhauer on the Character of the World, the Metaphysics of Will.*

Basavayya. D, "Effect of Bilingualism on Language Achievement, CIIL, Mysore, 1974.

Bear. R. M, "Factors Affectively the Success of College Freshmen", Journal of Applied Psychology: 21:517 - 523, Berger, IL and AR Sutker. (1928).

Bhadhuri, A (1971),A " Comparative Study of Certain Psychosocial Characteristics of the over and the under Achieves in Higher Secondary Schools., D.phi., psy, Cas U.

Bhart B. Dhar (2008) "Training and Development in Higher Education Academic Staff Challenges", A.P.H. Publishing Corporation.

Bhattacharya. I & Sharma .K, ' India in The Knowledge Economy – an Electronic Paradigm', International Journal of Education Management ,Vol.21, No.6, (2007).

Bhattach ... ınology-
As ... lucation
Int ...

Bose, P.K ... ; system
of ... pon it"
De ...

Chandra ... ing The
Le ... a', The
In¹ ... 07).

Chatterji ... Effection
ce: ... evement
Of ... Service
Ur ...

Chester .... _xecutive" Harvard university press.

Cholin.V.S, ' Study of The Application of Information Technology for Effective Access to Resource in Indian University Libraries', The Information International Information & Library Review 37 (3),(2005).

Choudhari, UP Jain , " Achievement of The School Children Calcutta: Psychometric Research and Service Unit, Indian Statistical Institute", (1975).

Clark, E. L, Iamily, Back Ground and college success, school and soc,. 1927.

CPSIA information can be obtained
at www.ICGtesting.com
Printed in the USA
BVHW060847140123
656278BV00010B/1265